FLORA OF TROPICAL EAST AFRICA

MELASTOMATACEAE

G. E. Wickens

Trees, shrubs or more commonly herbs, rarely climbers, sometimes epiphytes. Leaves opposite and decussate, rarely verticillate, sometimes anisophyllous, simple to entire, serrate, exstipulate, usually with a distinctive nervation of 2–8 strong basal nerves ± parallel to the midrib, rarely pinnately nerved. Inflorescences various, sometimes bracteate. Flowers ⚥, regular (androecium sometimes slightly zygomorphic), mostly 4–5(–6)-merous. Calyx tubular or campanulate, free or partially adnate to the ovary or sometimes connected to it by septa-like strands (no distinction is made here between the hypanthium and calyx proper); lobes valvate or connate and forming a calyptra-like head in bud, deciduous or persistent; various appendages may alternate with or arise from the calyx-lobes. Petals 4–5, free or rarely united at the base, imbricate or convolute, often magenta coloured. Stamens perigynous or epigynous, as many as the petals or more usually twice as many as the petals and in 2 whorls, the whorls equal to very unequal; anthers usually 2-thecous, introrse, basifixed, each anther dehiscing by a single pore or more rarely by a slit (*Memecylon*); anther-connectives often elongated and often either tubercled or spurred at the junction with the filament; filaments often geniculate, inflexed in bud, sometimes twisted. Ovary 1-locular with basal placentation (*Memecylon*) to several-locular with axile placentation; ovules usually numerous; style and stigma 1. Fruit a loculicidal capsule or a berry. Seeds small or minute, without endosperm and with a very small embryo; one cotyledon larger than the other.

Pantropical with about 200 genera and 4500 species, usually in areas of high rainfall, mainly in S. America, less numerous in Asia, Polynesia and Africa, very rare in the subtropics.

The genera of *Melastomataceae* are critical and the following key is not of gener a application outside the Flora area. A conspectus of subfamilies and tribes is given on p. 4. The subfamily *Memecyloideae* is sometimes given family status, e.g. Airy Shaw, Dict. Fl. Pl. & Ferns, ed. 7: 712 (1966) & ed. 8: 731 (1973).

Tibouchina viminea (D. Don) Cogn. (*T. urvilleana* (DC.) Cogn.), native of South America, is widely cultivated in the tropics as an ornamental shrub or small tree and is recorded from Kenya, S. Nyeri District, Ragati Forest Station, *Kokwaro* 2244 !, and from Tanzania, e.g. Amani, *Greenway* 3700 ! and Lushoto, *Semkiwa* 90 ! Among the native plants it would key to *Dissotis* subgen. *Dissotidendron*, but differs by the very large flowers (petals 2·5–4 cm. long), the silky pilose calyx (hairs 2–3 mm. long), the glandular-hairy filaments, etc. *Heterocentron subtriplinervum* (Link & Otto) A. Braun & Bouché, also native of tropical America, has recently been introduced into Nairobi gardens, e.g. *Gillett* 18538 ! & *Stewart* in *E.A.H.* 14010 ! It is easily recognized by the leaves, which are mostly on abbreviated lateral shoots, having a number of ascending pinnate nerves (not all meeting at or near the base); the delicate pink or white flowers are borne in cymose panicles.

Ovary 3–5-locular, with many ovules on axile placentas; fruit with numerous small seeds; flowers usually large and showy; anthers ovoid to linear, opening by an apical pore (to p. 4):

Plants with stems:

Ovary convex, sometimes with a rim, a crown of scales or bristles around the style-base, these structures not markedly accrescent:

Calyx with or without appendages, but never with a single long appendage from the back of each calyx-lobe; extension of anther-connective short or long, but if spurred behind then also with anterior tubercles or appendages:

Stamens with the connective usually extended below the anther (not or scarcely in *Osbeckia* and *Tristemma*), with anterior paired tubercles or appendages, very rarely with a projecting spur-like posterior appendage (then free part of connective long); seeds usually cochleate, i.e. coiled like a snail-shell (not in *Dichaetanthera*, scarcely in *Guyonia*):

Anthers ovoid or obovoid to shortly oblong, not attenuate, 1·5–4 times as long as broad; calyx ovoid to broadly urceolate-campanulate; stamens equal:

Shrub; appendages of connective slender, elongate, upcurved; ovary glabrous; intersepalar appendages caducous . . . . . . 1. **Dionychastrum**

Annual herbs; appendages of connective minute, tubercular:

Calyx without supplementary appendages; ovary glabrous; flowers single or in clusters of 2–3 . 2. **Guyonia**

Calyx with ligulate setose appendages alternating with the lobes; ovary with a crown of bristles; flowers up to 12 in heads . . . 3. **Antherotoma**

Anthers linear or linear-subulate, attenuate, 6 or more times as long as broad or if a little shorter (*Tristemma*) then calyx tubular-campanulate:

Seeds straight; trees, usually deciduous; flowers 4-merous; stamens similarly shaped, the antipetalous ones a little smaller . . . 5. **Dichaetanthera**

Seeds cochleate; herbs, shrubs or if trees (*Dissotis* subgen. *Dissotidendron*) then flowers 5-merous and stamens markedly dimorphic:

Fruit indehiscent; ovary largely adherent to the calyx; calyx glabrous or with 1–several discrete rings of bristles; stamens equal or subequal, with very short connective-extension and small appendages; flowers usually subtended by large bracts . 6. **Tristemma**

Fruit a capsule valvate within the calyx; ovary adhering for the greater part by 8–10 septa (most easily seen in bud when cavities occupied by inflexed anthers); calyx without discrete rings of bristles, sometimes glabrous:

Connective not or barely produced below the anther; stamens 10, equal; calyx-lobes persistent, with small linear setose intersepalar appendages . . 4. **Osbeckia**

Connective distinctly produced below the anther and if short other characters not as above:

Flowers 5-merous, solitary or capitate, surrounded by an involucre of large persistent bracts; calyx tubular-campanulate with persistent lobes and no supplementary appendages . . . 7. **Melastomastrum**

Flowers 4–5(–6)-merous, variously arranged but without an involucre of large persistent bracts; calyx with persistent or caducous lobes and usually with supplementary appendages* . 8. **Dissotis**

Stamens with the connective not extended below the anther (anthers basally attenuate on long stamens of *Dicellandra*), but produced into a well-developed pendent posterior and 2 upcurved anterior spur-like appendages; seeds straight (sometimes spurred):

Fruit a capsule; seeds obovoid with a distal lateral spur; flowers in terminal panicles, large (petals 13–18 mm.); main longitudinal nerves of leaf extending to the base and with conspicuous parallel transverse veins 14. **Dicellandra**

Fruit a berry; seeds semi-ovate, not spurred; flowers in axillary cymes, rather small (petals 4–6 mm.); main longitudinal nerves of leaf not all reaching the base, transverse veins inconspicuous . . . . . 15. **Medinilla**

Calyx-lobes small, with a longer subulate appendage from the back; connective only shortly extended below the anther, with a minute posterior appendage;

* If cultivated see *Tibouchina*, noted above.

leaves with the inner longitudinal nerves not quite reaching the base and with conspicuous parallel transverse veins; fruit a small berry; seeds straight . 16. **Clidemia**

Ovary depressed, the upper side ± flattened or concave, ± angular, surrounded by a rim-like crown of firm scales, the crown accrescent, slightly woody and persistent on the capsule, exserted well above the ± truncate calyx; stamens 10, equal or subequal; seeds straight*:

Flowers in umbels, long-pedicellate; connective produced at the base of the anther into a stout posterior spur . . . 9. **Gravesia**

Flowers in lax secund cymes, subsessile; connective shortly extended below the anther, with a small scale-like anterior appendage, the posterior appendage minute or lacking 10. **Calvoa**

Plants acaulous, the leaves and scapes arising from sometimes tuberous rhizomes:

Flowers in umbels, 2–6; anther-connective with 2 distinct anterior appendages; leaves subentire to minutely denticulate, shortly petiolate . . . . 11. **Primularia**

Flowers in scorpioid cymes (sometimes only 1–3); anther-connective with very reduced anterior appendages (appearing as tubercles or just a rim) and a small posterior appendage; leaves serrate, long-petiolate:

Stamens 8; anthers oblong-elliptic; ovary adnate to the calyx for most of its length; placentas short, restricted to the upper part of the ovary-axis . . . . . 12. **Cincinnobotrys**

Stamens 4 or 8; anthers oblong-ovate, attenuate to a narrow rostrate tip; ovary adherent to the calyx by 4–8 septa (easiest seen in bud); placentas extending the length of the ovary-axis . . . . 13. **Gravesiella**

Ovary 1-locular, with 5–20 ovules on a central placenta; fruit a 1–2-seeded berry; flowers small; anthers, including the thickened dorsal appendage, axe-shaped, medifixed, opening by longitudinal slits; trees or shrubs . . . 17. **Memecylon**

SYNOPSIS OF SUBFAMILIES AND TRIBES NATIVE TO THE FLORA AREA

A. Ovary 3–5-locular, with numerous axile ovules; fruit with numerous small seeds; flowers usually large and showy; anthers ovoid, oblong or linear, opening by an apical pore (subfamily *Melastomatoideae*).

B. Fruit a capsule, opening regularly by 4 or 5 valves (in *Guyonia* valvate within the membranous calyx, which tends to rupture irregularly), or

* Jacques-Félix has recently redefined *Dicellandra* to include W. African species which have these characters except that the crown is not accrescent. The single E. African species has a slightly concave ovary, but is easily distinguished from other E. African *Sonerileae* by the capsule, the unequal stamens and the spurred seeds.

rarely (*Tristemma*) capsule indehiscent; stamens equal, subequal or markedly differing.

C. Seeds sessile, cochleate (except *Dichaetanthera*); capsule terete or angular, apex convex; inflorescence paniculate, racemose, capitate or corymbose.

Tribe 1. **Osbeckieae** *DC.*, Prodr. 3: 127 (1828). Flowers 4–5-merous or very rarely 6–7-merous. Calyx-lobes distinct, often alternating with appendages. Stamens twice as many as the petals; anthers mostly linear, rostrate, acute or truncate, opening by a terminal pore; anther-connective sometimes produced, with a bilobed anterior appendage or paired appendages, rarely appendiculate posteriorly. Ovary usually with an apical crown of scales or bristles, rarely glabrous, 4–7-locular, usually adnate to the calyx-tube by septa (wholly adnate to the base of the calyx-tube in *Guyonia*). Fruit a capsule, rarely a berry. Seeds usually cochleate, rarely cuneiform. Distribution: Old World tropics. Genera 1–8.

C. Seeds funiculate, straight, raphe present; ovary and capsule 4–5-angled, apex of capsule concave; inflorescence umbelliform or a scorpioid cyme.

Tribe 2. **Sonerileae** (*Naud.*) *Triana* in Bull. Congr. Bot. Amsterdam 1865: 457 (1865). Flowers 4–5-merous (or 3-merous but not in the Flora area). Calyx-tube angular, rarely terete; lobes short, sometimes alternating with appendages or groups of hairs. Stamens twice as many as the petals or equal in number; anthers usually linear or ovate, acute, rostrate or rarely truncate, opening by 1 or rarely 2 apical pores; anther-connective not or shortly produced, with 2 anterior appendages or exappendiculate, and 1 short posterior spur or exappendiculate. Ovary 3–5-locular, adnate to the calyx-tube by septa or wholly adnate, apex widened, often with a crown of erect scales, usually glabrous. Capsule 3–5-angled, rarely terete. Seeds variable in shape and size, cuneate, obovoid, subreniform or irregular, straight or slightly curved, without wings, sometimes beaked. Distribution: Old World tropics. Genera 9–14.

B. Fruit baccate, rupturing irregularly; ovary usually adnate to the calyx-tube by septa, rarely wholly adnate.

Tribe 3. **Dissochaeteae** (*Naud.*) *Triana* in Bull. Congr. Bot. Amsterdam 1865: 457 (1865). Flowers 4–5(–6)-merous. Calyx-lobes insignificant or absent, rarely distinctly developed, not alternating with appendages. Stamens 4–30, usually twice as many as the petals or rarely equal in number, equal, subequal or unequal; anthers usually isomorphous, mostly elongate, acute or rarely rostrate or obtuse, opening by 1 or 2 terminal pores or, rarely, introrsely by 2 slits; connective usually produced; anterior spurs usually absent. Ovary 4–5–6-locular, adnate to the calyx-tube by septa or, rarely, wholly adnate, apex often swollen, sometimes provided with a short disk, glabrous or hairy. Fruit usually a ± cylindrical berry. Seeds straight or slightly curved, obovoid or irregular. Distribution: Old World tropics. Genera 15, 16.

A. Ovary 1-locular, completely adnate to the calyx-tube; ovules 2–20 on a central placenta; fruit a berry, 1–5-seeded; seeds large; flowers small; anther with appendage axe-shaped, medifixed, opening by longitudinal slits; trees or shrubs (subfamily *Memecyloideae*).

Tribe 4. **Memecyleae** *Cham.* in Linnaea 10: 217 (1835). Flowers 4–5-merous. Calyx-lobes usually insignificant. Stamens 8 or 16, equal or subequal; connective ending dorsally in a thick appendage at least as long as the thecae and provided with a dorsal sessile hollow gland. Ovary 1–8-locular, wholly adnate to the calyx-tube, apex depressed, glabrous. Distribution: pantropical. Genus 17.

1. DIONYCHASTRUM

A. & R. Fernandes in Bol. Soc. Brot., sér. 2, 30: 169, t. 2 (1956)

Small shrub. Flowers 5-merous. Calyx-tube broadly campanulate, ± densely setose; calyx-lobes oblong, persistent, a little shorter than the tube; intersepalar teeth caducous, setose. Petals obovate, bilobed, shortly clawed. Stamens 10, equal; anthers oblong; connective shortly produced below the anther, with 2 anterior upcurved appendages. Ovary 5-locular, basal part

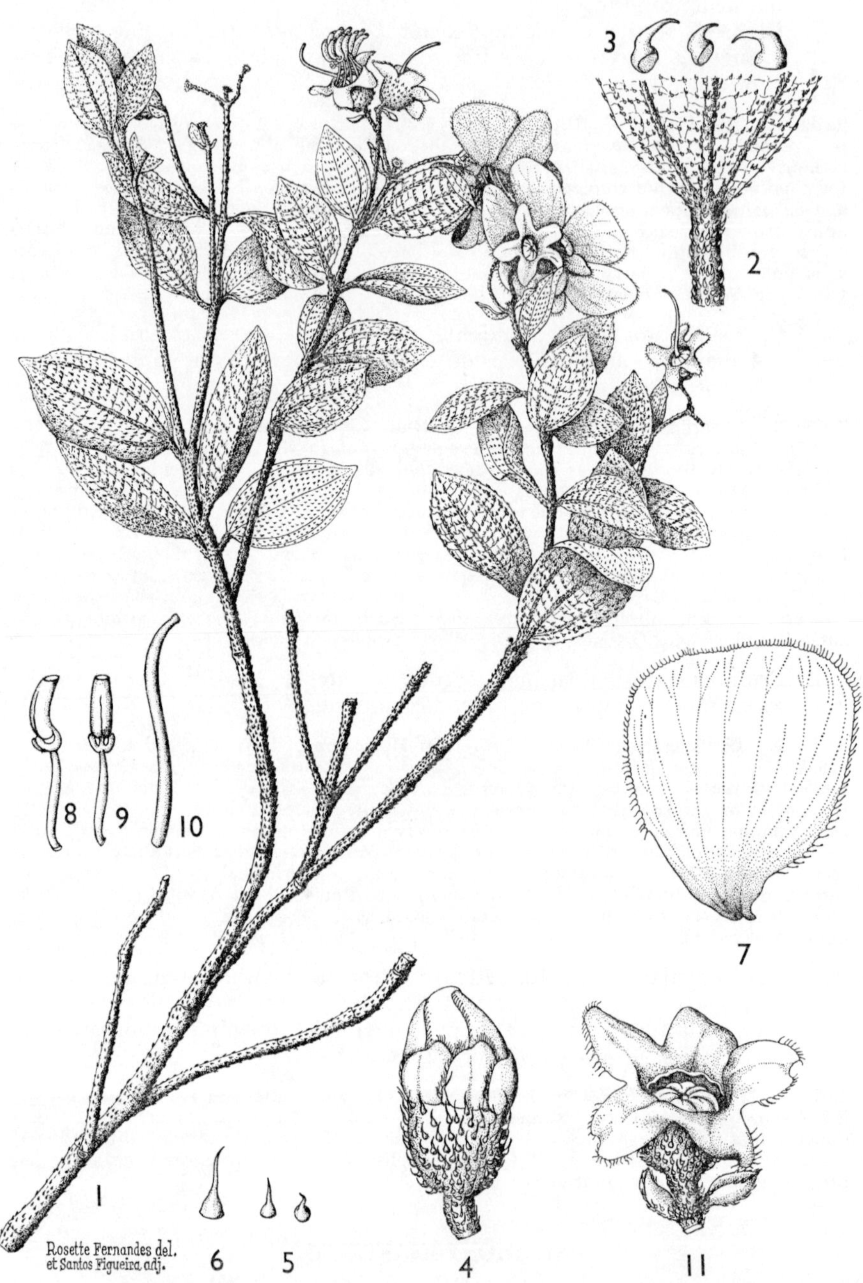

FIG. 1. *DIONYCHASTRUM SCHLIEBENII*—**1,** flowering branch, × 1; **2,** base of leaf, × 3; **3,** setae from petiole, × 10; **4,** flower bud, × 3; **5,** setae from base of calyx, × 6; **6,** intersepalar appendage, × 6; **7,** petal, × 3; **8,** stamen, side view, × 3; **9,** same, front view, × 3; **10,** style, × 3; **11,** fruiting calyx, with included capsule, × 3. All from *Schlieben* 3601. Reproduced by permission of the Editors of Boletim de Sociedade Broteriana.

adherent to the calyx-tube, subcylindrical, glabrous. Fruit a valvate capsule within the calyx. Seeds cochleate.

Monotypic; related to *Dionycha* Naud. from Madagascar.

D. schliebenii *A. & R. Fernandes* in Bol. Soc. Brot., sér. 2, 30: 170, t. 2 (1956). Type: Tanzania, Morogoro District, Mkambaku Mt., *Schlieben* 3601 (PRE, holo., COI!, EA, K!, LISC!, iso.)

Shrub 1 m. high; branches subterete, densely clothed with bulbous-based setae. Leaf-lamina elliptic, 1·5–2·5 cm. long, 0·8–1·5 cm. wide, apex subacute, base rounded or shortly cuneate, with numerous bulbous-based setae above, on the margins and on the venation beneath; main nerves 3–5, impressed above, prominent beneath; petiole 3–4 mm. long. Inflorescence terminal, 1–3-flowered; peduncle ± 1 cm. long; pedicel ± 2 mm. long, with 2 caducous ovate bracteoles 2·5 mm. long and 2 mm. wide. Calyx-tube 5 mm. long, 6 mm. in diameter, clothed in bulbous-based setae; lobes broadly oblong, obliquely truncate or emarginate, 4 mm. long, 3 mm. wide, margins ciliate, otherwise glabrous. Petals 12 mm. long, 10 mm. wide, pale rose, margins ciliate. Stamens 8 mm. long; anthers 3 mm. long; connective produced ± 1 mm., with 2 appendages ± 1 mm. long; filaments 4 mm. long. Ovary 3 mm. long, 2·5 mm. in diameter; style 11 mm. long. Fig. 1.

Tanzania. Morogoro District: Mkambaku Mt., 26 Feb. 1933, *Schlieben* 3601! & Magari peak above Mzinga, 6 Dec. 1970, *Pócs, Kondela & Nchimbi* 6300/F!
Distr. **T6**; not known elsewhere
Hab. Upland evergreen bushland; 2340–2640 m.

2. GUYONIA

Naud. in Ann. Sci. Nat., sér. 3, 14: 149, t. 6 (1850)

Afzeliella Gilg in E.M. 2: 4 (1898)

Creeping annual herbs. Flowers 1–3, terminal or axillary, 4–5–6-merous (varying even on the same plant). Calyx-tube ovoid; lobes narrow, persistent, without intersepalar appendages. Stamens equal; anthers ovoid to shortly oblong; connective produced below the anther, with 2 inconspicuous tubercular anterior appendages at the junction of the filament. Ovary adherent to the calyx-tube in the lower half, ovoid, glabrous; style sigmoid. Capsule valvate within the membranous calyx, which tends to rupture irregularly. Seeds only slightly cochleate, nearly turbinate, with a rounded wrinkled upper half.

Species 2, in tropical Africa.

G. ciliata *Hook. f.* in F.T.A. 2: 443 (1871); Cogn. in A. & C. DC., Monogr. Phan. 7: 337 (1891); Jacques-Félix in Bull. Soc. Bot. Fr. 97: 225 (1950); F.W.T.A., ed. 2, 1: 246 (1954); Jacques-Félix in Ic. Pl. Afr. 3, No. 50 (1955); A. & R. Fernandes in Mem. Soc. Brot. 11: 7 (1956) & in C.F.A. 4: 124 (1970). Type: Sierra Leone, *Afzelius* (?BM, holo.)

Slender annual herb; stem creeping with erect branches, rooting at the nodes, 4-angled, glabrous or appressed setose. Leaves decussate; lamina rhombic-ovate to broadly ovate, up to 3·2 cm. long and 2 cm. wide, apex subobtuse to acute, base shortly attenuate, membranous, crenulate, pinkish green beneath, sparsely setose above and on the nerves beneath; nerves 3–5, scarcely prominent beneath; petiole slender, up to 13 mm. long. Flowers 1–3, subtended by a caducous dentate-ciliate bract. Calyx-tube 3·5–4 mm. long, 3·5 mm. in diameter, sparsely pilose to glabrous; lobes linear-triangular,

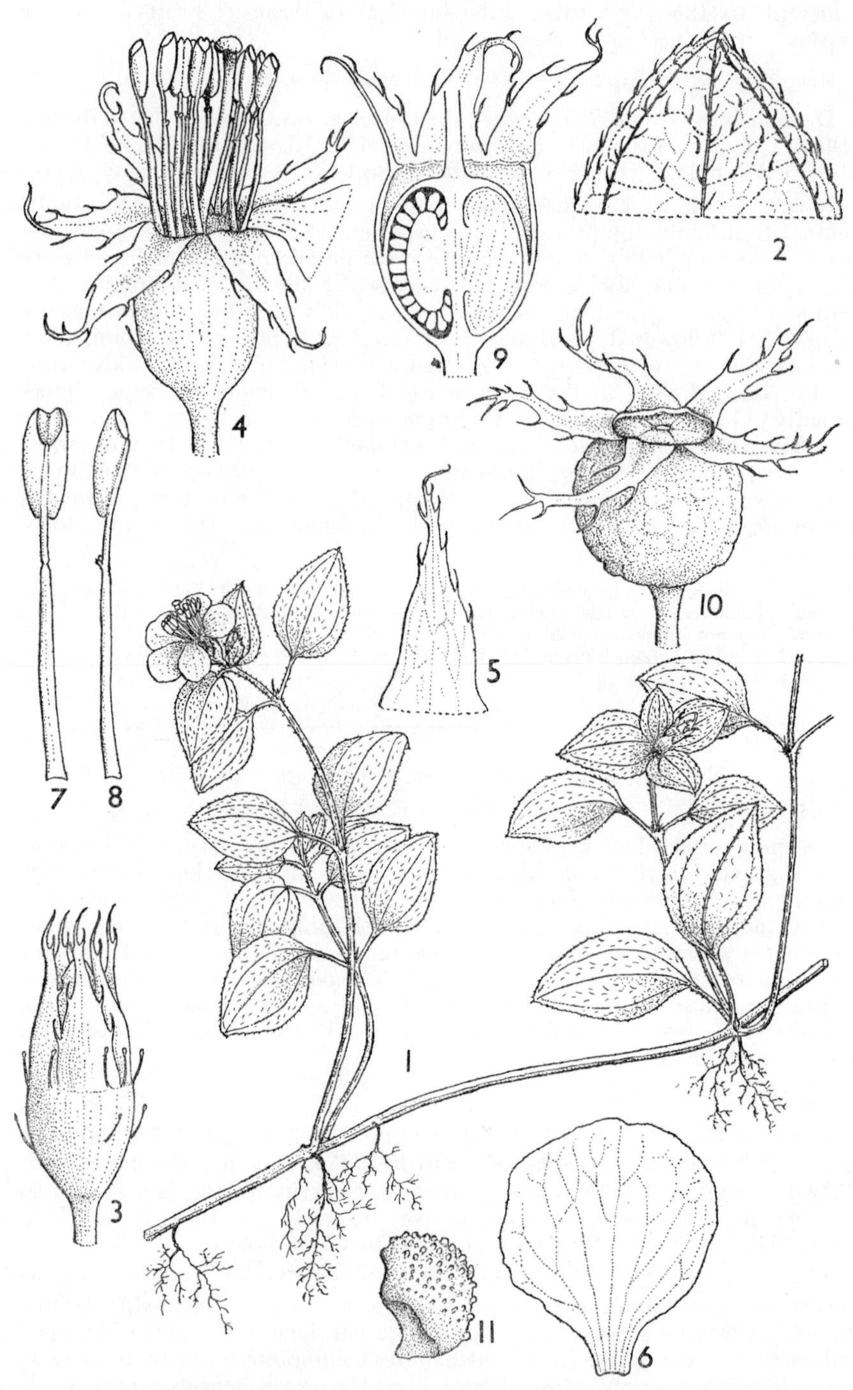

FIG. 2. *GUYONIA CILIATA*—**1,** habit, × 1; **2,** part of leaf, lower surface, × 4; **3,** flower bud, × 6; **4,** flower, with petals mostly removed, × 6; **5,** calyx-lobe, × 8; **6,** petal, × 4; **7, 8,** stamen, back and side views respectively, × 10; **9,** longitudinal section of flower, × 6; **10,** fruiting calyx, with included capsule, × 6; **11,** seed, × 20. 1–9, from *Purseglove* 1753; 10, 11, from *Burbridge* 153. Drawn by Mrs. M. E. Church.

5 mm. long, sparsely ciliate, persistent. Petals obovate, ± 8 mm. long, 4 mm. wide, delicate, pale pink. Stamens 5–5·5 mm. long; anthers 1–1·5 mm. long; connective ± 0·5–1 mm. long with 2 minute tubercles; filament 3–3·5 mm. long. Fruiting calyx 4–6 mm. in diameter, retaining the persistent lobes. Seeds 1·2–1·5 mm. long. Fig. 2.

UGANDA. Kigezi District: Kirima, Feb. 1949, *Purseglove* 2708!; Masaka District: Bugala I., Towa Forest, July 1945, *Purseglove* 1753! & 18 June 1951, *G. H. Wood* X.30!

TANZANIA. Bukoba District: near Bukoba, Aug. 1931, *Haarer* 2145!

DISTR. **U**2, 4; **T**1; forest regions of W. Africa from Guinée to Cameroun, Zaire and Angola

HAB. Epiphyte or ground herb in moist sites in rain-forest; 1050–1500 m.

SYN. *Afzeliella ciliata* (Hook. f.) Gilg in E.M. 2: 5 (1898); V.E. 3 (2): 744 (1921)
Guyonia intermedia Cogn. in De Wild. & Th. Dur., Pl. Thonner.: 30, t. 16 (1900). Type: Zaire, Ngali, *Thonner* 21 (BR, holo.)
Afzeliella intermedia (Cogn.) Gilg in Z.A.E.: 582 (1913)

NOTE. The second species, *G. tenella* Naud., on which the genus is based, occurs in Guinée and differs in being entirely glabrous. *G. ciliata* may prove to be no more than a setose form of *G. tenella*.

3. ANTHEROTOMA

(Naud.) Hook. f. in G.P. 1: 745 (1867)

Osbeckia L. sect. *Antherotoma* Naud. in Ann. Sci. Nat., sér. 3, 14: 55 (1850)

Erect annual. Flowers in heads, 4-merous. Calyx-tube ovoid-urceolate; lobes persistent, alternating with slender setose intersepalar appendages. Petals obovate, with an apical tuft of hairs. Stamens equal; anthers oblong-elliptic, truncate, with a large pore; connective produced below the anther, arched and with 2 tubercular anterior appendages at the junction of the filament. Ovary partly adherent to the calyx-tube, convex, with a small setose crown around the style-base. Capsule enclosed in the dry calyx-tube, valvate. Seeds small, cochleate.

Two species, one widespread in tropical Africa and Madagascar, the other in Cameroun. J. D. Hooker, in F.T.A. 2: 444 (1871), and more recently A. & R. Fernandes, in Garcia de Orta 2: 181 (1954), include as a second species, *Osbeckia decandra* (Sm.) DC. (*O. afzelii* Hook. f.), which is superficially similar but differs in having 5-merous flowers and anther-connectives without appendages. The generic limits between *Osbeckia, Antherotoma, Dissotis*, etc., are still subject to controversy.

A. naudinii *Hook. f.* in G.P. 1: 745 (1867) & in F.T.A. 2: 444 (1871); Triana in Trans. Linn. Soc. 28: 57, t. 4/43 (1871); Gilg in E.M. 2: 9, t. 1/f (1898); V.E. 3 (2): 746, fig. 317/D (1921); F.W.T.A., ed. 2, 1: 247 (1954); Jacques-Félix in Ic. Pl. Afr. 3, No. 51 (1955); A. & R. Fernandes in Mem. Soc. Brot. 11: 8 (1956) & in Bol. Soc. Brot., sér. 2, 34: 180 (1960) & in C.F.A. 4: 124 (1970). Types: Madagascar, *Bojer* & Comoros, Mayotte, *Boivin* 3418 (both K, syn.!, P, isosyn.)

Erect annual, rarely behaving as a biennial, (2·5–)4–35 cm. high; stem simple or branched, 4-angled, with appressed hairs on the angles, hairs somewhat spreading at the nodes. Leaves decussate, progressively larger towards the inflorescence, often widely spaced (internodes up to 22 cm.); lamina narrowly ovate to oblong-ovate, up to 3·8 cm. long and 1·4 cm. wide, apex obtuse, base cuneate to subtruncate, sparsely strigose above and on the nerves beneath; venation impressed above, prominent beneath, with 3 or 5 ascending nerves; petiole 1–5 mm. long, pilose. Inflorescence terminal or axillary, subtended by 1–2 pairs of leaves, capitate, up to ± 12-flowered;

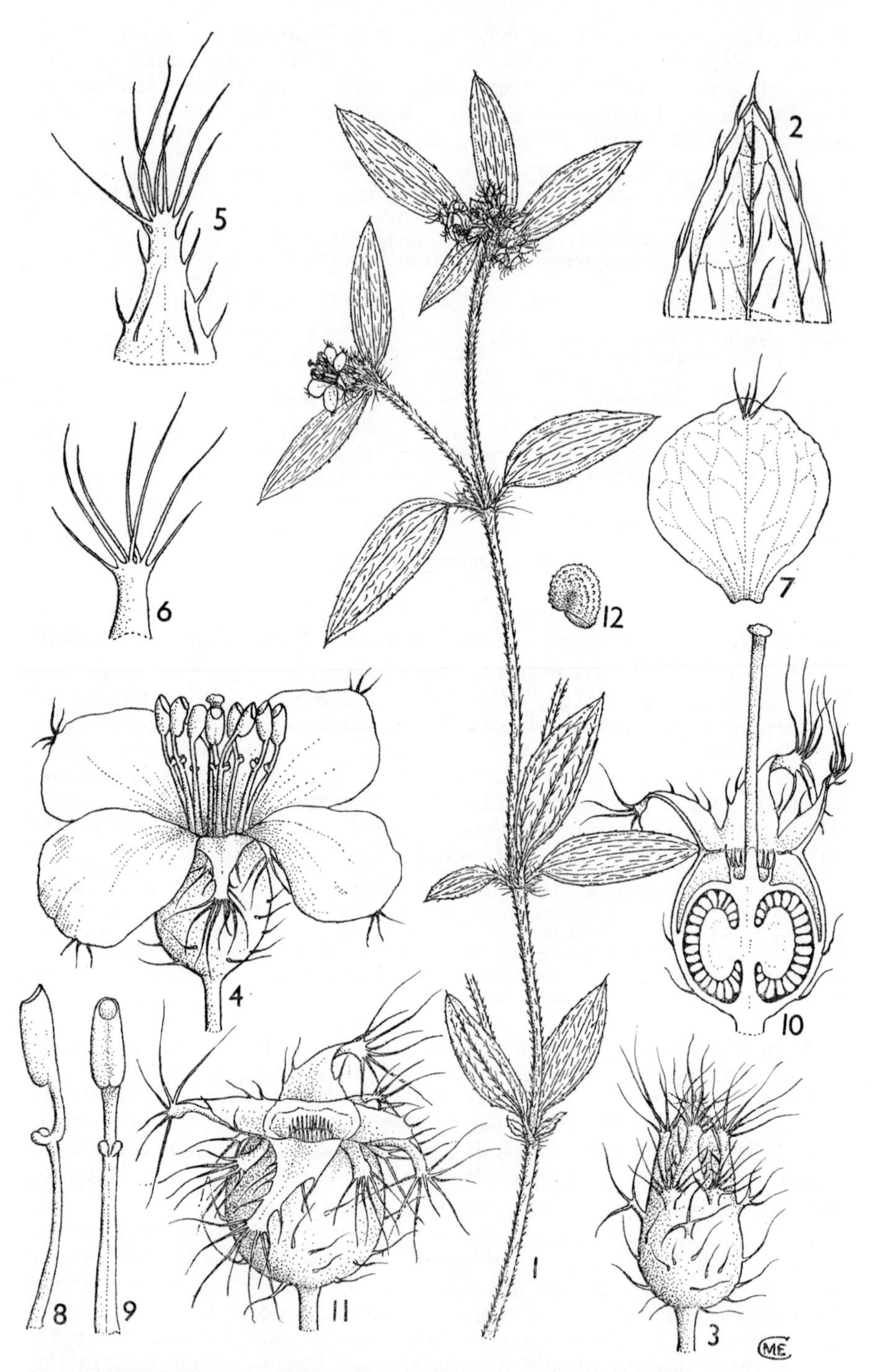

FIG. 3. *ANTHEROTOMA NAUDINII*—**1,** flowering branch, × 1; **2,** part of leaf, lower surface, × 6; **3,** flower bud, × 8; **4,** flower, × 6; **5,** calyx-lobe, × 10; **6,** intersepalar appendage, × 10; **7,** petal, × 6; **8, 9,** stamen, side and front views respectively, × 12; **10,** longitudinal section of flower, × 8; **11,** fruiting calyx, with included capsule, × 8; **12,** seed, × 20. All from *Milne-Redhead & Taylor* 9493. Drawn by Mrs. M. E. Church.

pedicels up to 3 mm. long. Calyx 2·5 mm. long, 2 mm. in diameter, sparsely hairy; lobes narrowly triangular, persistent, 1 mm. long, with reddish stellately arranged bristles at the apex; intersepalar appendages ligulate, with reddish stellately arranged bristles at the apex. Petals 4 mm. long, 3·5 mm. wide, pale mauve, with a tuft of hairs at the apex. Stamens 3·5 mm. long; anthers 0·7 mm. long; free part of connective ± 0·5 mm. long, arched, 2-tubercled at the base. Capsule rounded, 2·5–3 mm. long. Seeds 0·4–0·5 mm. long, very finely papillose. Fig. 3.

UGANDA. W. Nile District: Aringo County, Chei Hill, Sept. 1937, *Eggeling* 3410!; Ankole District: Igara, Mar. 1939, *Purseglove* 632!; Teso District: 24 km. Soroti–Moroto, 13 Oct. 1952, *Verdcourt* 822!

KENYA. W. Suk/Elgeyo Districts: Cherangani Hills, 12 Oct. 1961, *Bogdan* 5280!; Uasin Gishu District: Ol Dane Sapuk, 28 July 1951, *Greenway* 8553!; N. Kavirondo District: Kakamega Forest, 15 Oct. 1953, *Drummond & Hemsley* 4758!

TANZANIA. Bukoba District: Maruku, *Panayotis* 76!; Buha District: Gombe Stream Reserve, Mkenke stream valley, 28 Mar. 1964, *Pirozynski* 616!; Songea District: Matengo Hills, Mpapa, 25 May 1956, *Milne-Redhead & Taylor* 10442!

DISTR. **U**1–4; **K**?2, 3, 5; **T**1, 3–5, 7, 8; widespread in the forest margins and relic forest areas from Guinée across to Ethiopia and south to Angola and the Transvaal, also Madagascar, where it is believed to have been introduced (Perrier de la Bâthie, Fl. Madag. 153, Melastom.: 7 (1951))

HAB. Short grassland, usually in damp places, rock crevices and pavements, also by lakes, rivers and ditches; 300–2150 m.

SYN. *Osbeckia antherotoma* Naud. in Ann. Sci. Nat., sér. 3, 13: t. 6/10 (1849) & 14: 56 (1850); Taub. in P.O.A. C: 295 (1895). Type: as for the species

Antherotoma antherotoma (Naud.) Krasser in E. & P., Pf. 3, 7: 154 (1893), *nom. illegit.*

Dissotis kundelungensis De Wild. in Ann. Mus. Congo, Bot., sér. 4, 2: 117 (1913). Type: Zaire, Katanga, Kundelungu, *Kassner* 2731 (BR, holo., BM, K, iso.!)

4. OSBECKIA

L., Sp. Pl.: 345 (1753) & Gen. Pl., ed. 5: 162 (1754), emend. A. & R. Fernandes in Bol. Soc. Brot., sér. 2, 28: 75 (1954) & in Garcia de Orta 2: 188 (1954)

Herbs, subshrubs or shrubs, usually erect, hispid, strigose or scabrid. Leaves 3–7-nerved, entire or serrulate. Inflorescence capitate, racemose or panicled. Flowers 4–5-merous. Calyx ovoid, urceolate or subglobose, usually produced beyond the ovary, rarely glabrous, usually hispid, bristly or with scales bearing pectinate or stellately arranged bristles, sometimes with such appendages alternating with the calyx-lobes. Petals obovate, rose or purple. Stamens equal; anthers linear-subulate, incurved or sigmoid; connective not or only shortly produced below the anther, without appendages or with 2 anterior tubercles. Ovary adherent to the calyx, usually by 8–10 septa, 4–5-locular, with a crown of bristles at the apex. Capsule enclosed in the dry calyx-tube. Seeds minute, cochleate.

About 60 species in Asia, Australia and Madagascar, with 5 species in Africa.

A. & R. Fernandes (*loc. cit.*, 1954) believe that the genus *Dissotis* has been derived from *Osbeckia* and that the similarities observed between species belonging to the two genera in Africa are not due to parallel evolution. They believe that species of *Dissotis* are still capable of producing forms with the primitive staminal characteristics of *Osbeckia*. They have accordingly transferred Sect. *Pseudodissotis* Cogn. (in A. & C. DC., Monogr. Phan. 7: 331 (1891)) from *Osbeckia* to *Dissotis*, with the exception of the Asiatic species *O. cochinchinensis* Cogn. and *O. papuana* Cogn.

O. congolensis *Buettn.* in Verh. Bot. Ver. Brandenb. 31: 95 (1889); Cogn. in A. & C. DC., Monogr. Phan. 7: 314 (1891); Krasser in E. & P., Pf. 3, 7: 156 (1893); Gilg in E.M. 2: 6 (1898); V.E. 3 (2): 744 (1921); Jacques-Félix in Ic. Pl. Afr. 3, No. 52 (1955); A. & R. Fernandes in Mem. Soc. Brot. 11: 67

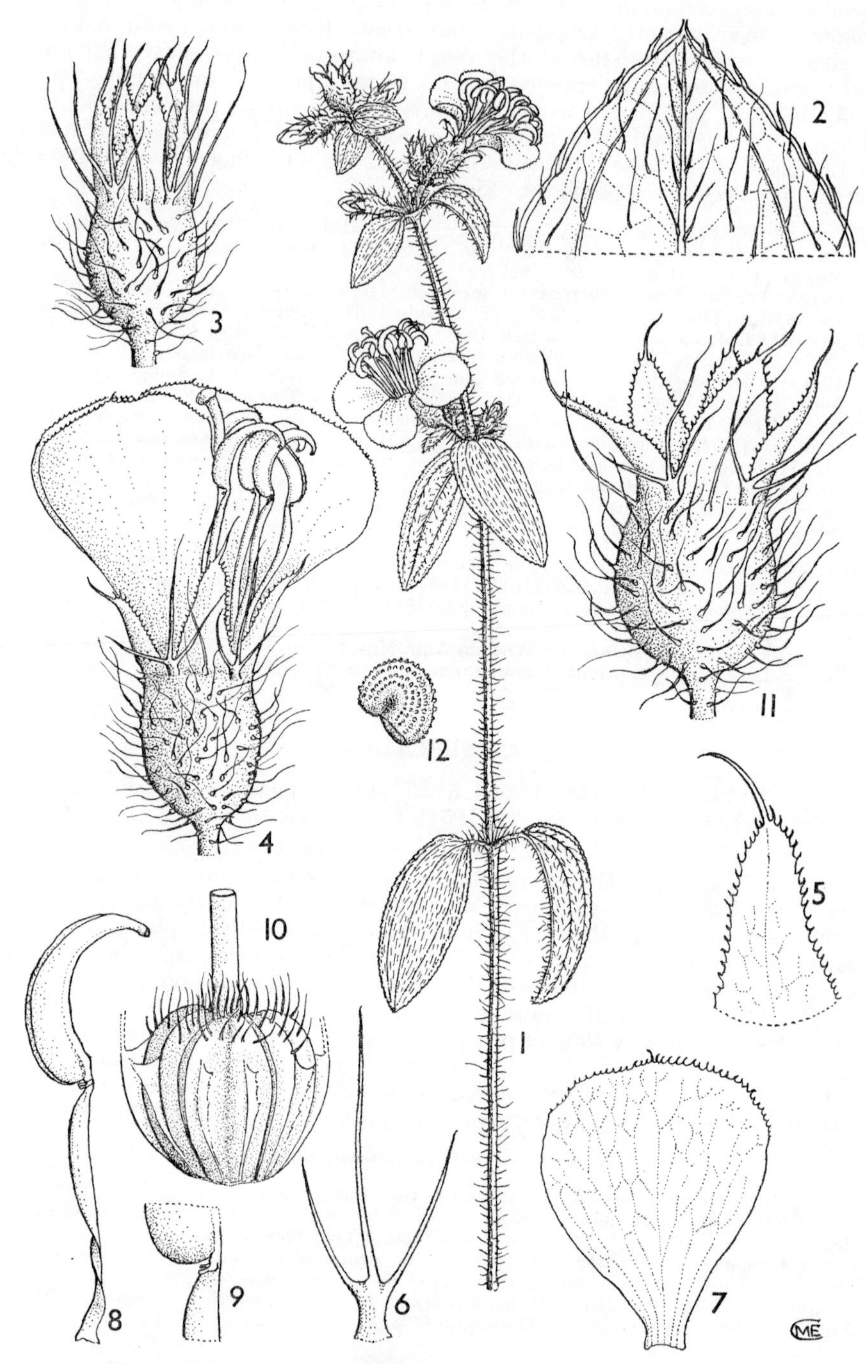

FIG. 4. *OSBECKIA CONGOLENSIS*—**1**, flowering branch, × 1; **2**, part of leaf, lower surface, × 6; **3**, flower bud, × 4; **4**, flower, with one petal removed, × 4; **5**, calyx-lobe, × 8; **6**, intersepalar appendage, × 10; **7**, petal, × 4; **8**, stamen, side view, × 6; **9**, detail of same showing connective-appendages, × 10; **10**, ovary, × 8; **11**, fruiting calyx, × 6; **12**, seed, × 20. 1–11, from *Lind* 2329; 12, from *Haswell* 81. Drawn by Mrs. M. E. Church.

(1956) & in C.F.A. 4: 129 (1970). Type: Zaire, Kibaka, *Buettner* 24 (B holo.†)

Annual herb, branched, erect, up to 60 cm. high; stem square, slightly winged, thinly long-setose, more densely setose at the nodes. Leaves decussate, shortly petiolate to subsessile; lamina oblong-elliptic, up to 5 cm. long and 1·5 cm. wide, apex obtuse to subacute, base cuneate to rounded, thinly long-setose above and on the nerves beneath; main nerves 3–5, slightly impressed above, subprominent beneath; petiole up to 5 mm. long. Inflorescence of terminal and axillary contracted cymes forming leafy false panicles. Flowers 5-merous, shortly pedicellate. Calyx-tube campanulate, 4–6 mm. long, ± densely long-setose; lobes oblong-triangular, 3 mm. long, persistent, margins ciliate, alternating with small linear appendages bearing long bristles. Petals ± 7 mm. long, 5 mm. wide, pinkish mauve, upper edge shortly ciliate. Stamens 10 mm. long; anthers 4 mm. long, recurved towards the apex; free part of connective very short, with 2 anterior protuberances (and sometimes 2 minute posterior ones); filaments twisted, 4–6 mm. long. Ovary ovoid, 3 mm. long, 2·5 mm. in diameter, lower half adherent to the calyx by septa, upper half free, setulose and with a crown of bristles at the apex surmounting the style-base; style sigmoid. Capsule ovoid, ± 5 mm. long. Seeds 0·6 mm. long, finely echinulate. Fig. 4.

UGANDA. Masaka District; Sese Is., Bugoma, June 1925, *Maitland* 784! & near Lake Nabugabo, Lake Nyarya, 22 Feb. 1958, *Lind* 2329!
TANZANIA. Bukoba District: Bugandika, Sept. 1931, *Haarer* 2188!; Tabora District: near Tabora, Uyogo, Nyumbe, 5 June 1913, *Braun* in *Herb. Amani* 5408!
DISTR. U4; T1, 4; W. Africa from Guinée to the Ivory Coast, also in Gabon, Zaire, Burundi, Zambia and Angola
HAB. Grassland, moist places, not common; ± 1200 m.

SYN. *O. congolensis* Buettn. var. *robustior* Buettn. in Verh. Bot. Ver. Brandenb. 31: 95 (1889); Cogn. in A. & C. DC., Monogr. Phan. 7: 314, 1177 (1891). Type: Zaire, Bolobo to Lukolela, *Buettner* 25 (B, holo. †)

5. DICHAETANTHERA

Endl., Gen.: 1215 (1839); Jacques-Félix in Bull. Soc. Bot. Fr. 102: 37 (1955)

Sakersia Hook. f. in G.P. 1: 757 (1867) & in F.T.A. 2: 458 (1871)
Barbeyastrum Cogn. in A. & C. DC., Monogr. Phan. 7: 376 (1891)

Trees or shrubs, usually deciduous. Inflorescence of cymose corymbs or panicles. Flowers 4-merous. Calyx campanulate; lobes absent or short and exposing the corolla in bud, persistent in fruit. Petals red, rarely white. Stamens similar in shape, but the antipetalous ones smaller; anthers linear-subulate; connective elongated below the anther, with 2 anterior appendages at the junction of the filament. Ovary adnate to the calyx-tube by 8 septa, apex convex, usually setose, the bristles sometimes forming a ring. Seeds ± cuneiform, sometimes shortly curved at one end and sometimes (not in E. Africa) ± cochleate at maturity; hilum lateral.

Species 34; 7 species in Africa, the remainder in Madagascar.

Calyx glabrous, 7–8 mm. long . . . . . . 1. *D. verdcourtii*
Calyx setose:
  Calyx 6 mm. long; calyx-lobes shallowly triangular, 1·5 mm. long. . . . . 2. *D. corymbosa*
  Calyx 9 mm. long; calyx-lobes ovate, 5 mm. long . 3. *D. erici-rosenii*

1. **D. verdcourtii** *A. & R. Fernandes* in Bol. Soc. Brot., sér. 2, 43: 297, t. 5–7 (1969). Type: Tanzania, Kigoma District, Uvinza–Mpanda, *Verdcourt* 3435 (EA, holo. !, COI, K, iso. !)

Tree up to 7–5 m. high, flowering before the leaves; bark corky; branches 4-angled, internodes of young branches densely strigose; nodes swollen, long-setose. Leaf-lamina ovate or broadly elliptic, 6–13 cm. long, 3·5–7·5 cm. wide, apex acute, base rounded or ± cuneate, strigose on both surfaces; midrib and the 2 pairs of basal longitudinal nerves impressed above, subprominent beneath. Inflorescence a lax terminal panicle, ± 20-flowered; panicle-axis sparsely setose below, glabrous towards the apex; pedicels 4–6 mm. long, glabrous. Calyx-tube campanulate, 7–8 mm. long, 5–6 mm. in diameter, glabrous; lobes ovate, 3 mm. long, glabrous except for the ciliate margins. Petals oblong-obovate or ovate, 22 mm. long, 12–14 mm. wide, purple. Anthers 11–12 mm. long; free part of connective curved, 12–14 mm. long; anterior appendages 5 mm. long; filament 10 mm. long. Ovary glabrous, without apical setae. Fruiting calyx accrescent, 9 mm. long, 8 mm. in diameter; capsule exserted.

TANZANIA. Kigoma District: 42 km. S. of Uvinza, 30 Aug. 1950, *Bullock* 3259! & 42 km. Uvinza–Mpanda, 23 Nov. 1962, *Verdcourt* 3435! & Kasakati, Sept. 1965, *Suzuki* B.17!

DISTR. **T4**; not known elsewhere

HAB. *Brachystegia* woodland, sometimes in rocky gorges; 1500–1600 m.

2. **D. corymbosa** (*Cogn.*) *Jacques-Félix* in Bull. Soc. Bot. Fr. 102: 38 (1955); A. & R. Fernandes in Mem. Soc. Brot. 11: 45 (1956); F.F.N.R.: 305 (1962); A. & R. Fernandes in C.F.A. 4: 160 (1970). Type: Zaire, Bateke–Osika, *Brazza* (P, holo.)

Tree 9–15 m. high; branches 4-angled, internodes of young branches densely strigose; nodes swollen, long-setose. Leaf-lamina ovate-elliptic, 6–11·5 cm. long, 2·7–4·5 cm. wide, apex acute, base rounded, strigose above, densely strigose beneath, with longer setae on the nerves; midrib and 2 pairs of basal longitudinal nerves impressed above, prominent beneath; lateral nerves subprominent beneath; petiole 1–1·7 cm. long. Inflorescence a lax terminal panicle, up to 40–50-flowered; axis strigose; pedicels 2–3 mm. long, strigose. Calyx-tube cupular, 6 mm. long, 5·5 mm. in diameter, strigose; lobes shallowly triangular, 1·5 mm. long, lightly strigose. Petals oblong-obovate, 10–12 mm. long, 8 mm. wide, rose-pink. Stamens with antipetalous ones somewhat smaller; anthers linear-subulate, 6 and 8 mm. long; connectives produced 3 and 4 mm.; anterior appendages 3 mm. long; filaments 10 or 12–13 mm. long. Ovary with a dense apical crown of setae. Fruiting calyx accrescent, 8 mm. long, 6 mm. in diameter, the capsule barely exserted.

UGANDA. Ankole District: Kalinzu Forest, Kyanga Camp, Dec. 1931, *Gibson* 6/M.A in *F.D.* 385!; Kigezi District: Bufumbira, Nyamagana, June 1947, *Purseglove* 2456! & Kigezi, Mar. 1933, *Sanford* in *A. S. Thomas* 1211!

DISTR. **U2**; Cameroun, Zaire, Zambia and Angola

HAB. Upland rain-forest; 1450–1950 m.

SYN. *Barbeyastrum corymbosum* Cogn. in A. & C. DC., Monogr. Phan. 7: 376 (1891); Gilg in E.M. 2: 23, t. 1/G (1898); V.E. 3 (2): 753, fig. 317/E (1921)

Sakersia laurentii Cogn. in Ann. Mus. Congo, Bot., sér. 2, 1: 23 (1899); I.T.U., ed. 2: 171 (1952). Types: Zaire, Bumba, 1896, *Laurent* & without locality, *Dewèvre* (both BR, syn.)

S. corymbosa (Cogn.) Jacques-Félix in Bull. I.F.A.N. 15: 1001 (1953)

3. **D. erici-rosenii** (*R. E. Fries*) *A. & R. Fernandes* in Bol. Soc. Brot., sér. 2, 30: 181, t. 16, 17 (1956) & in 34: 69 (1960); F.F.N.R.: 306, fig. 53 (1962). Type: Zambia, R. Kalungwishi, *R. E. Fries* 1154 (UPS, holo.)

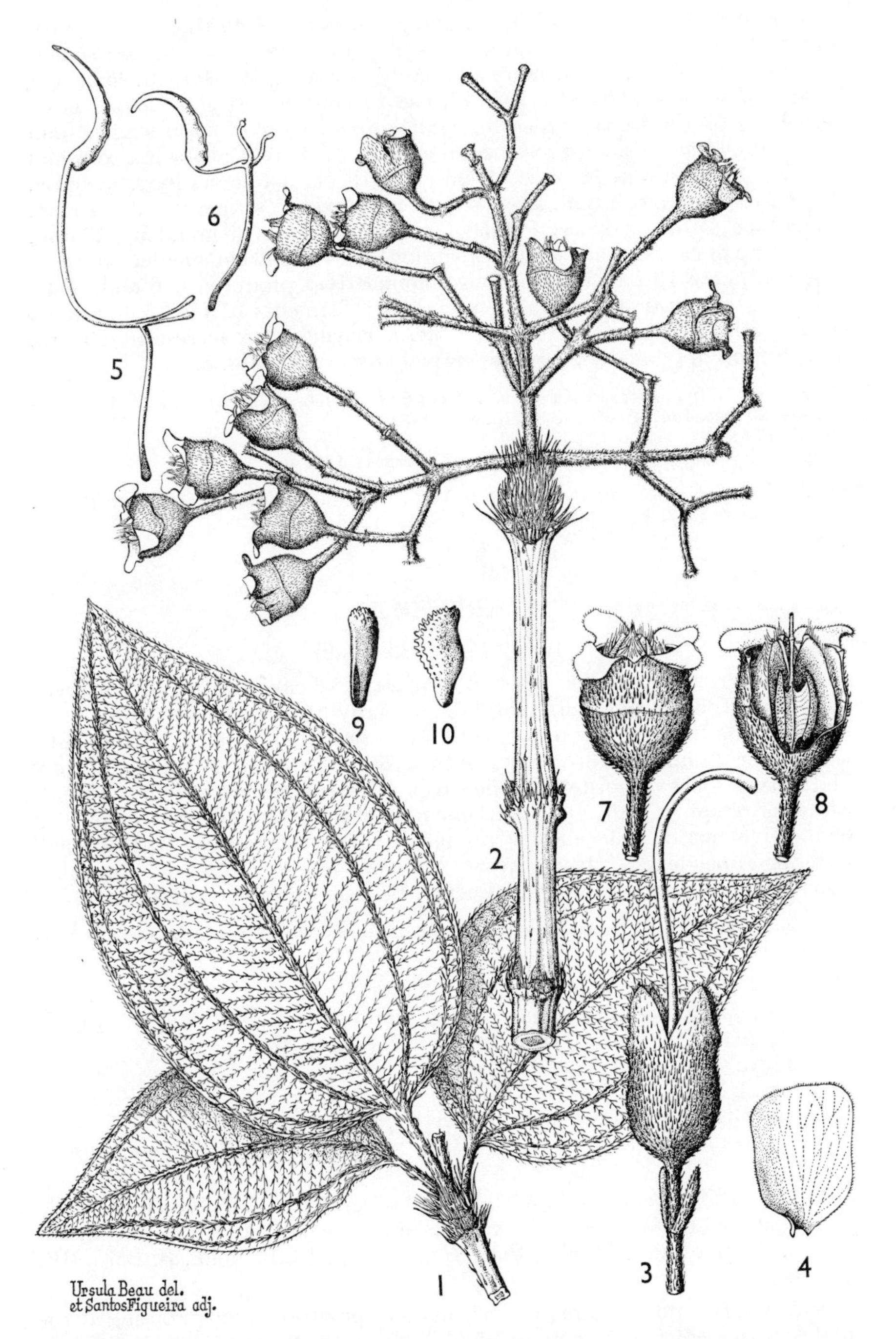

FIG. 5. *DICHAETANTHERA ERICI-ROSENII*—**1,** tip of leafy branch, × 1; **2,** infructescence, × 1; **3,** flower, after fall of petals and stamens, × 2; **4,** petal, × 1; **5, 6,** large and small stamens respectively, × 2; **7,** fruiting calyx, × 2; **8,** same, opened to show dehisced capsule, × 2; **9, 10,** seed, front and side views respectively, approx × 20. 1, 2, 7–10, from *Angus* 797; 3–6, from *R. G. Robertson* 170. Reproduced by permission of the Editors of Boletim da Sociedade Broteriana.

Tree up to 4·5(–8 in Zambia) m. high, flowering before the leaves; bark corky; branches 4-angled; internodes of young branches strigose; nodes swollen, long-setose. Leaf-lamina broadly ovate, up to 8·5 cm. long and 7 cm. wide, apex subacute, base obtuse to cordate, strigose above, more densely so on the nerves beneath; midrib and 2 pairs of basal longitudinal nerves impressed above, prominent beneath. Inflorescence a lax terminal panicle, ± 30-flowered; axis strigose; pedicels 3–5 mm. long, strigose. Calyx-tube cupular, 9 mm. long, 7 mm. in diameter, strigose; lobes ovate, 5 mm. long, lightly strigose. Petals oblong-trapeziform, 30 mm. long, 22 mm. wide, deep pink. Stamens with antipetalous ones somewhat smaller; anthers linear-subulate 10 and 12 mm. long; connectives produced 4–6 and 14–20 mm. long; anterior appendages 3 mm. long; filaments 12 and 15 mm. long. Ovary with a dense apical crown of setae. Fruiting calyx accrescent, 10 mm. long, 9 mm. in diameter; capsule scarcely exserted. Fig. 5.

TANZANIA. Ufipa District: Kawa [Kara] R., 2 Oct. 1956, *Richards* 6347! & Kalambo Falls, 20 Oct. 1967, *Simon & Williamson* 1134!
DISTR. **T4**; Zambia
HAB. *Brachystegia* woodland, sometimes in gorges; 1200 m.

SYN. *Dissotis erici-rosenii* R. E. Fries in Wiss. Ergebn. Schwed. Rhod.-Kongo-Exped. 1911–1912, 1: 179, t. 13/15–18 (1914)

6. TRISTEMMA

Juss., Gen.: 329 (1789)

Shrubs or woody herbs, erect or prostrate. Leaves entire, 5–7-nerved, petiolate. Inflorescence terminal, 1–3– or many-flowered; flowers 5-merous, sessile, usually enclosed by several large persistent bracts. Calyx-tube tubular or campanulate, usually with 1 or more rings of bristles, rarely glabrous; lobes persistent, reflexed; intersepalar appendages absent. Stamens equal or subequal; anthers narrowly oblong to linear-subulate; connective not or only shortly produced below the anther, with 2 small anterior appendages. Ovary mostly adherent to the calyx-tube, 5-locular, apex usually setose. Fruit indehiscent. Seeds cochleate.

About 16 species, all in Africa with 1 extending across to Madagascar and the Mascarene Islands.

Flowers 1–3; bracts small, less than 4 mm. long, not enveloping the inflorescence; calyx glabrous . . . 1. *T. leiocalyx*
Flowers many, or if few then bracts enveloping the calyx:
 Calyx glabrous . . . . 2. *T. schliebenii*
 Calyx with 1–3 rings of hairs or bristles, rings sometimes incomplete . . . . 3. *T. mauritianum*

1. **T. leiocalyx** *Cogn.* in A. & C. DC., Monogr. Phan. 7: 1179 (1891); Gilg in E.M. 2: 24 (1898); A. & R. Fernandes in Bol. Soc. Brot., sér. 2, 34: 68, 191 (1960). Type: Zaire, Stanley Pool, *Hens* sér. B, 13 (G, holo., B, BM!, BR, K!, iso.)

Soft woody shrub forming a dense mass of prostrate stems, rooting at the nodes, with vertical shoots up to 1·5 m. high; stem 4-angled, lightly strigose. Leaf-lamina ovate-elliptic, up to 10·5 cm. long and 5·5 cm. wide, shortly acuminate, base shortly attenuate to subtruncate, shortly strigose on both surfaces; nerves 5, subprominent on both surfaces; petiole up to 3 cm. long. Inflorescence terminal or axillary; flowers 1–3, subsessile; bracts broadly ovate, acute, 2–4 mm. long, membranaceous, strigose. Calyx-tube tubular-campanulate, 6–7 mm. long, 3–3·5 mm. in diameter, glabrous; lobes narrowly triangular, 2–3 mm. long, margins shortly ciliate. Petals obovate, 6–8 mm. long, 5–6 mm. wide, white or pink. Stamens equal, 5 mm. long; anthers 2 mm. long; connective produced 0.6 mm., with 2 anterior tubercles 0·5 mm. long; filaments 2–3 mm. long. Ovary ovoid, ± 5 mm. long, apex setose; style ± 6 mm. long. Fig. 6/13, p. 18.

Uganda. Kigezi District: Ishasha Gorge, May 1950, *Purseglove* 3429!; Masaka District: NW. side of Lake Nabugabo, 9 Oct. 1953, *Drummond & Hemsley* 4703!; Mengo District: 10 km. N. of Entebbe, Ziku Forest, 16 Mar. 1950, *Dawkins* 595!

Distr. U2, 4; Cameroun, Zaire and Sudan

Hab. Marshy clearings in lowland rain-forest, riverine and swamp forest; 1140–1320 m.

Syn. *T. roseum* Gilg in E.M. 2: 24, t. 1/J (1898). Types: Zaire, Khor Assika, Mar. 1870, *Schweinfurth* 3161 (B, syn. †) & 3323 (B, syn. †, K, isosyn.!) & Khor Kussumaba, Apr. 1870, *Schweinfurth* 3656 (B, syn. †)

Tetraphyllaster rosaceum Gilg in E. & P. Pf., Nachtr. 1: 266 (1897). Type: Cameroun Mt., collector not cited (B, holo. †)

2. **T. schliebenii** *Markgraf* in N.B.G.B. 14: 107 (1938); A. & R. Fernandes in Mem. Soc. Brot. 11: 42, t. 6 (1956). Type: Tanzania, Mafia I., *Schlieben* 2656 (B, holo. †, BR, P, Z, iso.)

Softly woody shrub up to 1·5 m. high; stem 4-angled, spreading or retrorsely hispid, densely hispid at the nodes. Leaf-lamina ovate-elliptic, up to 16 cm. long and 7·5 cm. wide, shortly acuminate, base shortly attenuate to rounded, chartaceous, margin inconspicuously toothed, thinly long-pilose on both surfaces; nerves 5(–7 *fide* Markgraf), impressed above, subprominent beneath; petiole up to 2·5 cm. long. Inflorescence terminal; peduncle less than 5 mm. long; flowers 1–6, subsessile; bracts boat-shaped, 1·5–2·5 cm. long, 1 cm. wide, scarious, margins and nerves strigose. Calyx-tube tubular-campanulate, 10 mm. long, 6 mm. in diameter, glabrous; lobes narrowly triangular-acuminate, 6–8 mm. long, 2 mm. wide, margins setose-ciliate. Petals oblong-obovate, 20 mm. long, 7 mm. wide, pale purple to magenta, glabrous. Stamens subequal; anthers 7–8(–10 *fide* Markgraf) mm. long, incurved; connective 0·2–0·7(–1) mm. long, with 2 anterior tubercles; filaments ± 10 mm. long. Ovary ovoid, 8 mm. long, glabrous to sparsely setose near the apex; style 1·8 cm. long. Fig. 6/12, p. 18.

Tanzania. Rufiji District: Mafia I., 3 Apr. 1933, *Wallace* 758! & 10 Sept. 1937, *Greenway* 5237!

Distr. T6; known only from Mafia I.

Hab. Cyperaceous swamps; 5 m.

3. **T. mauritianum** *J. F. Gmel.*, Syst. Pl. 2: 693 (1791). Type: Mauritius, *Commerson* (P–JU, holo.)

Shrubby herb up to 2 m. high; branches 4-angled, shortly winged, hispid. Leaf-lamina elliptic to elliptic-ovate, up to 20 cm. long and 14·5 cm. wide, apex acute or acuminate, base attenuate to rounded, thinly strigose on both surfaces; nerves 5–7, impressed above, prominent beneath; venation sub-prominent beneath; petiole 1–4 cm. long. Inflorescence terminal; peduncle 2–4 mm. long; flowers ± 2–15; outer bracts leaf-like, slightly exceeding the flowers; inner bracts obovate, equalling the calyx-tube, coriaceous, glabrous except for the ciliate margins. Calyx-tube tubular-campanulate, 10–11 mm. long, 5–6 mm. in diameter, with 1–2(–5) complete annular rings of bristles, sometimes with an additional incomplete ring, or with 1–2 incomplete rings only; calyx-lobes triangular, 5–8 mm. long, 2·5–4 mm. wide, margins ciliate. Petals obovate, up to 15 mm. long, 12 mm. wide,

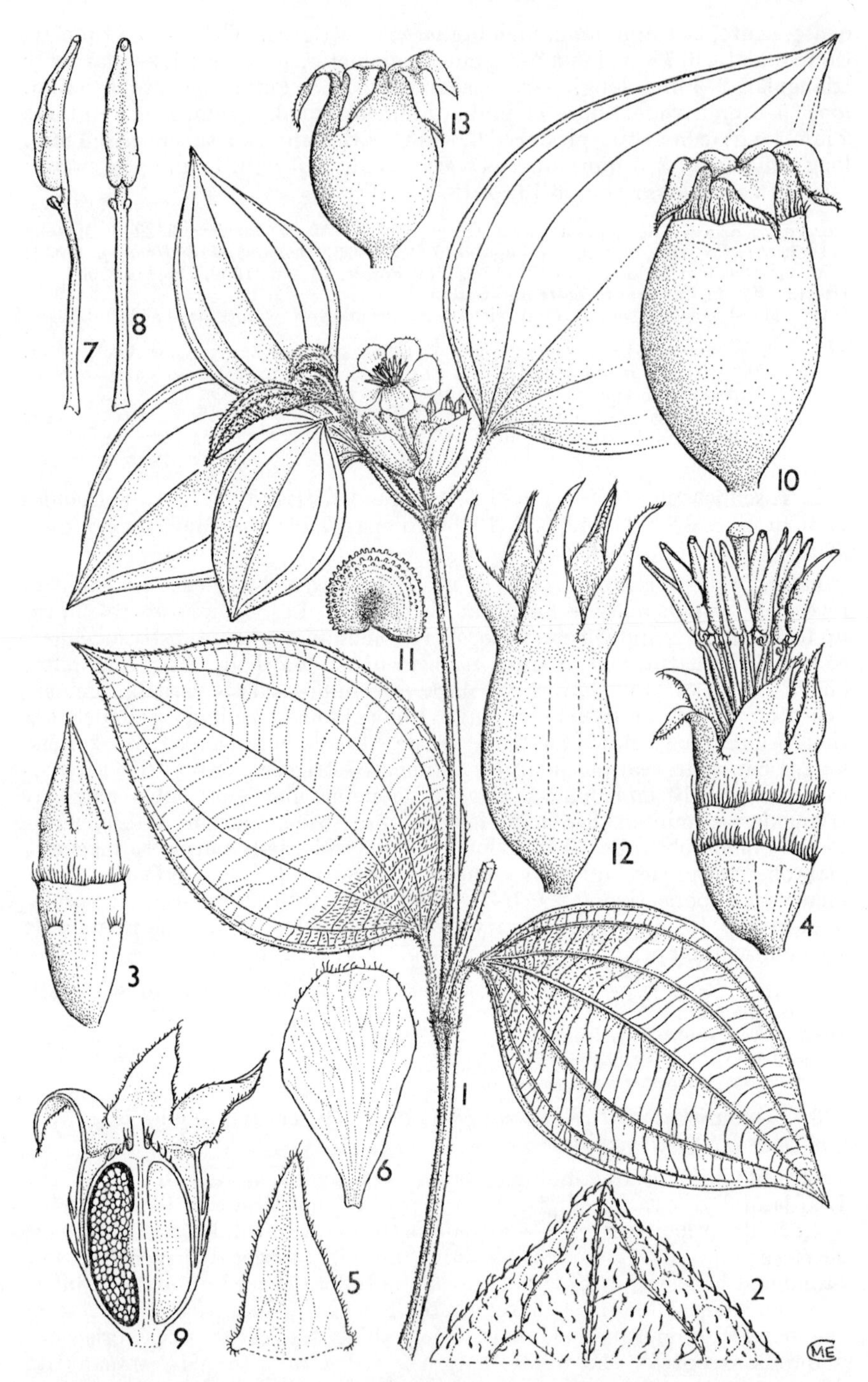

FIG. 6. *TRISTEMMA MAURITIANUM*—**1,** flowering branch, × $\frac{2}{3}$; **2,** part of leaf, lower surface, × 4; **3,** flower bud, × 4; **4,** flower, with petals removed, × 4; **5,** calyx-lobe, × 6; **6,** petal, × 3; **7, 8,** stamen, side and front views respectively, × 6; **9,** longitudinal section of flower, × 4; **10,** fruiting calyx, × 3; **11,** seed, × 20. *T. SCHLIEBENII*—**12,** fruiting calyx, × 3. *T. LEIOCALYX*—**13,** fruiting calyx, × 3. 1–9, from *Symes* 418; 10, 11, from *Osmaston* 2777; 12, from *Wallace* 758; 13, from *Dawkins* 595. Drawn by Mrs. M. E. Church.

pink to pale mauve. Stamens equal; anthers 3 mm. long; connective 0·2 mm. long with 2 anterior tubercles; filaments 3·5 mm. long. Ovary ovoid, ± 8 mm. long, apex setose; style up to 1·5 mm. long. Fig. 6/1–11.

UGANDA. Kigezi District: Amahenge, Aug. 1949, *Purseglove* 3052!; Mbale District: Budadiri, Jan. 1932, *Chandler* 510!; Mengo District: Namanve Forest, 12 Feb. 1963, *Tallantire & J. M. Lee* 636!

KENYA. N. Kavirondo District: Bukura, 23 Mar. 1944, *M. D. Graham* in *A.D.* 62! Kwale District: Shimba Hills, Giriama Point, 30 Oct. 1968, *Glover & Estes* 1153!

TANZANIA. Mwanza District: Ukerewe I. (or the area Mwanza to Musoma), 12 Oct. 1929, *Conrads* 578!; Buha District: Kakombe, 7 July 1959, *Newbould & Harley* 4266!; Mpanda District: Pasagulu–Musenabantu, 10 Aug. 1959, *Harley* 9287!

DISTR. **U**1–4; **K**5, 7; **T**1, 3, 4, 6, 8; **P**; widespread in tropical Africa from Senegal eastwards to Ethiopia and southwards to Angola, Rhodesia and Mozambique, also in Madagascar and the Mascarene Is.

HAB. Marshy clearings in rain-forest, swampy riverine forest; 775–1950 m.

SYN. *T. virusanum* Juss., Gen.: 329 (1789), *nom. inval.*; Humbert in Fl. Madag. 153, Melastom.: 4, fig. 1/1–5 (1951). Based on *Commerson* from Mauritius

T. incompletum R. Br. in Tuckey, Narr. Exped. Zaire: 435 (1818) & as separate: 16 (1818); Gilg in E.M. 2: 25 (1898); V.E. 3 (2): 754 (1921); F.P.S. 1: 195 (1950); F.W.T.A., ed. 2, 1: 250 (1954); A. & R. Fernandes in Mem. Soc. Brot. 11: 42, 94 (1956) & in Bol. Soc. Brot., sér. 2, 34: 191 (1960); F.P.U., ed. 2: 93 (1971). Type: Zaire, R. Congo, *Smith* (BM, holo.!, K, iso.!)

Melastoma albiflorum G. Don, Gen. Syst. 2: 764 (1832). Type: Sierra Leone, *Don* (BM, holo.!)

Tristemma schumacheri Guill. & Perr. in Fl. Seneg. Tent. 1: 311 (1833); Naud. in Ann. Sci. Nat., sér. 3, 13: 298, t. 6/10 (1850); Hook. f. in F.T.A. 2: 44 (1871); Triana in Trans. Linn. Soc. 28: 56 (1871); Cogn. in A. & C. DC., Monogr. Phan. 7: 461, 1180 (1891), incl. var. *grandifolia* Cogn.; Taub. in P.O.A. C: 295 (1895). Types: Gambia, Albreda, 1827, *Perrottet* & Senegal, Casamance, 1829, *Perrottet* (both P, syn.)

T. albiflorum (G. Don) Benth. in Hook., Niger Fl.: 353 (1849); Cogn. in A. & C. DC., Monogr. Phan. 7: 362 (1891)

T. fruticulosum Gilg in E.M. 2: 26, t. 1/M (1898); T.T.C.L.: 313 (1949). Type: Tanzania, Bukoba, *Stuhlmann* 1002 (B, holo. †)

T. grandifolium (Cogn.) Gilg in E.M. 2: 26, t. 1/N (1898); V.E. 3 (2): 755 (1921); A. & R. Fernandes in Bol. Soc. Brot., sér. 2, 34: 192 (1960) & in C.F.A. 4: 128 (1970). Type: Angola, Golungo Alto, *Welwitsch* 900 (LISU, lecto., BM!, COI, K!, P, iso.)

T. incompletum R. Br. var. *grandifolium* (Cogn.) Hiern, Cat. Afr. Pl. Welw. 1: 364 (1898)

T. grandifolium (Cogn.) Gilg var. *congolanum* De Wild. in Ann. Mus. Congo, Bot., sér. 5, 2: 329 (1908). Types: Zaire, Kimuenza, *Gillet* 1979 & Kasongo, *Dewèvre* 938a & Eala, *Laurent* 1926 and a number of other syntypes (all BR, syn.)

T. kassneranum Kraenzlin in Viert. Nat. Ges. Zürich 76: 148 (1931). Type: Zaire, Semliki R., *Kassner* 3094 (BM, K, iso.!)

T. acuminatum A. & R. Fernandes in Bol. Soc. Brot., sér. 2, 30: 184, t. 21–23 (1956) & in Mem. Soc. Brot. 11: 43, 95 (1956) & in Bol. Soc. Brot., sér. 2, 34: 192 (1960). Type: Tanzania, Lushoto District, Monga, *Greenway* 3683 (PRE, holo., EA, K!, iso.)

NOTE. The ± continuous variation found in this species makes the recognition of infraspecific taxa untenable.

7. MELASTOMASTRUM

Naud. in Ann. Sci. Nat., sér. 3, 13: 296 (1850); A. & R. Fernandes in Garcia de Orta 2: 278 (1954)

Shrubs or woody herbs. Flowers 5-merous, surrounded by an involucre of bracts. Calyx tubular-campanulate; lobes persistent. Petals obovate. Stamens 10, usually unequal; anthers linear-subulate; long stamens with arcuate extension of the connective and 2 anterior lobed appendages; short stamens with shorter less arcuate connective and with 2 anterior tubercular appendages; rarely all stamens similar. Ovary adhering to the calyx-tube by 10 septa, 5-locular, with a crown of bristles at the apex. Capsule enclosed in the dry calyx-tube. Seeds minute, cochleate.

Species 6, in tropical Africa.

Flowers several in sessile bracteate heads; base of flower with long (1·5–4 mm.) bristles . . . 1. *M. capitatum*
Flowers 1(–2), invested by bracts, the inflorescence so formed subsessile to shortly pedunculate; base of flower shortly setose . . . . 2. *M. segregatum*

1. **M. capitatum** (*Vahl*) *A. & R. Fernandes* in Garcia de Orta 2: 278 (1954) & in Mem. Soc. Brot. 11: 69 (1956); Keay in F.W.T.A., ed. 2, 1: 761 (1958); A. & R. Fernandes in C.F.A. 4: 130 (1970). Type: West Africa, ? Sierra Leone, collector not known (C (Herb. Schumacher), holo., K, photo.!)*

Woody herb or shrub up to 2 m. high; stems robust, erect or weak and flexuous, sparsely scabrid to densely strigose. Leaf-lamina ovate, up to 13 cm. long and 5 cm. wide, apex acute, base cuneate to shortly attenuate, strigose on both surfaces; nerves 5, impressed above, subprominent beneath; petiole up to 1·6 cm. long. Inflorescence of 2 or more flowers enclosed in an involucre of bracts, sessile; outer bracts leafy; inner bracts persistent, oblong, acute, membranous with tufts of long hairs towards the base. Calyx ± 11 mm. long, 5 mm. in diameter, crimson, glabrous except for a few bristles at the base; lobes lanceolate, 7 mm. long, 2·5 mm. wide, acuminate, margins sparsely ciliate. Petals broadly obovate, 20 mm. long, 15 mm. wide, mauve. Long stamens 22 mm. long, anthers 9 mm. long, mauve, free part of connective 4 mm. long with 2 lobes 1–2 mm. long; short stamens 12 mm. long, anthers 5 mm. long, yellow, free part of connective 0·5 mm. long with 2 lobes 1 mm. long. Ovary 5 mm. long, glabrous except for the apical crown of bristles; style ± 12 mm. long. Fig. 7/12.

UGANDA. W. Nile District: Mt. Otzi, Oct. 1959, *E. M. Scott* in *E.A.H.* 11800!; Kigezi District; Ishasha gorge, Nov. 1946, *Purseglove* 2272!; Busoga District: Musisi ridge, July 1936, *Sangster* 157!
TANZANIA. Bukoba, Aug. 1931, *Haarer* 2192!; Buha District: Kasekela Reserve, 17 Nov. 1962, *Verdcourt* 3334!; Mpanda District: Mahali Mts., Musuma R., 30 Sept. 1958, *Newbould & Jefford* 2787!
DISTR. **U**1–4; **T**1, 4; Senegal to the Sudan and south to Angola and Zambia
HAB. Edges of lowland rain-forest and open *Uapaca* woodland; 750–1350 m.

SYN. *Melastoma capitata* Vahl, Eclog. Pl. Amer.: 45 (1797)
M. capitata G. Don, Gen. Syst. 2: 764 (1832), *nom. illegit.* Type: Sierra Leone, *G. Don* (BM, holo.!)
Tristemma erectum Guill. & Perr. in Fl. Seneg. Tent. 1: 312 (1833). Type: Senegal, Kounoun, *Perrottet* (P, holo.)
Heterotis capitata Benth. in Hook., Niger Fl.: 352 (1849), *nom. illegit.*, based on *Melastoma capitata* G. Don, *non* Vahl
Melastomastrum erectum (Guill. & Perr.) Naud. in Ann. Sci. Nat., sér. 3, 13: 296, t. 5/4 (1850)
Tristemma capitatum (Vahl) Triana in Trans. Linn. Soc. 28: 56, t. 24/41d (1871)
Dissotis capitata (Benth.) Hook. f. in F.T.A. 2: 449 (1871); Cogn. in A. & C. DC., Monogr. Phan. 7: 365 (1891); Taub. in P.O.A. C: 295 (1895); Gilg in E.M. 2: 13 (1898); V.E. 3 (2): 747 (1921), *nom. illegit.*
D. petiolata Hook. f. in F.T.A. 2: 448 (1871); Cogn. in A. & C. DC., Monogr. Phan. 7: 363 (1891); Taub. in P.O.A. C: 295 (1895); Gilg in E.M. 2: 12 (1898);

* Vahl in his original description cites a plant said to have been collected by Schumacher in the West Indies, but Schumacher never went to the West Indies. The herbarium sheet however is annotated by Schumacher " D. Banks, Dryander ". It is believed that the original specimen may have been grown by Dryander at Kew from seed obtained from an unknown collector in West Africa and the plant obtained by Schumacher when he studied botany under Banks in 1788–89.

FIG. 7. *MELASTOMASTRUM SEGREGATUM*—**1,** flowering branch, $\times \frac{2}{3}$; **2,** part of leaf, lower surface, × 6; **3,** flower bud, × 2; **4,** calyx-lobe, × 4; **5,** petal, × 1; **6, 7,** large and small stamens respectively, × 3; **8,** longitudinal section of flower, × 4; **9,** fruiting branchlet, $\times \frac{2}{3}$; **10,** fruiting calyx, × 3; **11,** seed, × 20. *M. CAPITATUM*—**12,** fruiting branchlet, $\times \frac{2}{3}$. 1, from *Symes* 465; 2–11, from *Drummond & Hemsley* 4700; 12, from *Maitland* 1298. Drawn by Mrs. M. E. Church.

V.E. 3 (2): 747 (1921); F.P.S. 1: 192 (1950). Type: Uganda, W. Nile District, Madi, *Grant* (K, holo.!)
D. erecta (Guill. & Perr.) F. W. Andr., F.P.S. 1: 192, fig. 110 (1950); F.W.T.A., ed. 2, 1: 259 (1954)

NOTE. A. & R. Fernandes, in Mem. Soc. Brot. 11: 15 (1956) & Bol. Soc. Brot., sér. 2, 43: 285 (1969), divide the species into three varieties differing principally in the indumentum. The East African material is mostly referable to var. *capitatum*, but plants from more exposed sites approach var. *barteri* (Hook. f.) A. & R. Fernandes, which may in fact be based on no more than an ecotypic form of the species.

2. **M. segregatum** (*Benth.*) *A. & R. Fernandes* in Mem. Soc. Brot. 11: 12 & 68, t. 1 (1956) & in C.F.A. 4: 130 (1970). Types: Nigeria, R. Nun, *Vogel* 12 & Lokoja, *Ansell* & Nupe, *Barter* (all K, syn.!)

Shrub up to 3 m. high; stem strigose. Leaf-lamina narrowly elliptic to narrowly oblong-obovate, up to 12 cm. long and 4 cm. wide, apex acute, base rounded to truncate, shortly strigose on both surfaces; nerves 5, impressed above, prominent beneath; petiole up to 1·5 cm. long. Inflorescence 1(–2)-flowered, shortly pedunculate, each flower enclosed in an involucre of bracts; outer 2 bracts ± foliaceous, inner 4 bracts persistent, obovate, 9 mm. long, 5 mm. wide, truncate, scarious, glabrous. Calyx-tube 11 mm. long, 5 mm. in diameter, glabrous except for a few bristles at the base; lobes lanceolate, 5 mm. long, 3 mm. wide, margins ciliate. Petals broadly obovate, 20 mm. long, 16 mm. wide, pink. Stamens usually unequal; long stamens 25 mm. long, anthers 9 mm. long, mauve; free part of connective 6 mm. long with 2 lobes 1·5 mm. long; short stamens 15 mm. long, anthers 8 mm. long, yellow, free part of connective 0·7 mm. long with 2 lobes 1 mm. long; sometimes (forma *osbeckioides*) stamens all similar. Ovary 7 mm. long, glabrous except for the apical crown of bristles; style 20 mm. long. Fig. 7/1–11.

UGANDA. Masaka District: Bugala I., Kalangala, 25 Feb. 1945, *Greenway & Thomas* 7186! & NW. side of Lake Nabugabo, 9 Oct. 1953, *Drummond & Hemsley* 4700! & 5 km. N. of Lake Nabugabo, 25 Apr. 1971, *Lye* 5989!

TANZANIA. Bukoba District: Bukoba, Aug. 1931, *Haarer* 2099! & 16 Apr. 1948, *Ford* 350! & Bushasha, 1935, *Gillman* 324!; Pemba I., Pandani–Kinazini, 19 Feb. 1929, *Greenway* 1492!

DISTR. **U4**; **T1**; **P**; Nigeria, Cameroun, Zaire, Zambia, Rhodesia, Angola and South West Africa

HAB. Swamp and riverine forest; 1120–1170 m. inland, 15–30 m. on Pemba

SYN. *Heterotis segregata* Benth. in Hook., Niger Fl.: 350 (1849)
Dissotis segregata (Benth.) Hook. f. in F.T.A. 2: 448 (1871); Cogn. in A. & C. DC., Monogr. Phan. 7: 363 (1891); Taub. in P.O.A. C: 295 (1895); Gilg in E.M. 2: 12 (1898); V.E. 3 (2): 747 (1921); T.T.C.L.: 310 (1949); F.W.T.A., ed. 2, 1: 259 (1954)
Tristemma segregatum (Benth.) Triana in Trans. Linn. Soc. 28: 56, t. 4/41c (1871)
Dissotis minor Gilg in E.M. 2: 12, t. 2/C (1898); V.E. 3 (2): 747 (1921). Type: Tanzania, Bukoba District, Bumbiri [Bumbinde] I., *Stuhlmann* 3583 (B, holo. †)
Melastomastrum schlechteri A. & R. Fernandes in Bol. Soc. Brot., sér. 2, 29: 48, t. 2 (1955). Type: Cameroun, Ngoko, *Schlechter* 12782 (PRE, holo., BR, K!, P, iso.)

8. DISSOTIS

Benth. in Hook., Niger Fl.: 346 (1849); A. & R. Fernandes in Bol. Soc. Brot., sér. 2, 43: 285 (1969), *nom. conserv.*

Kadali Adans., Fam. 2: 234 (1763)
Kadalia Raf., Sylva Tell.: 101 (1838), *homon. orthogr.*
Dupineta Raf., Sylva Tell.: 101 (1838)
Hedusa Raf., Sylva Tell.: 101 (1838)

Heterotis Benth. in Hook., Niger Fl.: 347 (1849), pro parte
Argyrella Naud. in Ann. Sci. Nat., sér. 3, 13: 300 (1850)
Osbeckiastrum Naud. in Ann. Sci. Nat., sér. 3, 14: 118 (1850)
Lepidanthemum Klotzsch in Peters, Reise Mossamb., Bot. 1: 64 (1861)
Lignieria A. Chev., Expl. Bot. Afr. Occ.: 279 (1920)
[*Osbeckia* sensu auct. mult., pro parte, *non* L.]

Herbs, less often shrubs or small trees. Inflorescence of terminal panicles, rarely the flowers solitary; flowers 4–5(–6)-merous. Calyx-tube campanulate to ovoid, glabrous or variously hairy, with or without pectinate scales; lobes persistent or caducous; intersepalar appendages often present. Petals purple or violet. Stamens usually markedly unequal, sometimes subequal in "osbeckioid" forms; connective long-produced at the base of the large stamens with 2-tuberculate, 2-lobed or 2-spurred appendage at the junction with the filament; connective scarcely produced in the short stamens with 2-tuberculate, 2-lobed or 2-spurred appendage. Ovary 4–5-locular, adherent to the calyx-tube by 8–10 septa or partially free, apex setose. Capsule enclosed by the calyx, coriaceous. Seeds numerous, minute, cochleate.

About 140 species in tropical and southern Africa. A synopsis of subgenera and sections in the Flora area is given on p. 29. The structure and appendages of the calyx of most species are illustrated in figs. 8–12, p. 28–33.

Small trees or shrubs, the bark of branchlets peeling and flowers appearing before the leaves:
- Calyx-tube glabrous . . . . 9. *D. caloneura*
- Calyx-tube pubescent:
  - Calyx-tube covered by short (0·5 mm.), bulbous-based, upward-curving setae (fig. 9/8); leaf-lamina broadly ovate, base cordate . . 14. *D. bussei*
  - Calyx-tube and leaves not as above:
    - Indumentum of calyx-tube sparse, consisting of small simple setae mixed with setose appendages and sometimes with glandular setae (fig. 9/4, 5); leaf-lamina elliptic-lanceolate to elliptic, base truncate to subcordate . . . . 12. *D. melleri*
    - Indumentum of calyx-tube ± dense, at least in lower half:
      - Calyx-tube pilose on lower half, the upper half glabrous or sparsely pilose (fig. 9/6); leaf-lamina elliptic-ovate, base cuneate to subcordate . . . . . . 10. *D. arborescens*
      - Calyx-tube densely covered with glandular-setose tufts (fig. 9/7); leaf-lamina elliptic, base subcordate . . . . 11. *D. glandulicalyx*

Herbs, shrubs or small trees, if shrubs or small trees then other characters not as above:
- Flowers 4-merous; calyx-lobes persistent; inflorescence ± densely capitate; bracts small, persistent; erect annual or perennial herbs:
  - Indumentum on stem appressed:
    - Indumentum very sparse; intersepalar appendages bristle-like (fig. 8/1); stem slender, ± 0·5 mm. in diameter; leaves elliptic-lanceolate, subserrate, less than 1 cm. long 1. *D. tisserantii*
    - Indumentum ± dense; intersepalar ligulate appendages setose (fig. 8/2, 3):

Indumentum not concealing the stem; stem robust, usually more than 1 mm. in diameter; leaves narrowly elliptic, elliptic-lanceolate to linear-lanceolate, more than 1·5 cm. long . . . . 2. *D. debilis*

Indumentum dense, concealing the stem; inflorescence ± concealed by the leaves 3. *D. phaeotricha*

Indumentum on stem spreading; intersepalar appendages setose* . . . . . . 3. *D. phaeotricha*

Flowers 5-merous or if rarely 4-merous then calyx-lobes caducous:

Stem winged; woody herbs:

Calyx-tube 8–14 mm. long, densely silvery sericeous, hairs ± obscuring the small scale-like appendages:

Calyx-tube 8–9 mm. long, 5·5 mm. in diameter, with hairs 3–7 mm. long, fig. 12/4 (**T4**) . . . . . . 28. *D. pterocaulos*

Calyx-tube 8–14 mm. long, 9–13 mm. in diameter, with hairs 1–3 mm. long, fig. 12/5 (**T8**) . . . . . . 29. *D. formosa*

Calyx-tube 7–8·5 mm. long, 7 mm. in diameter, densely appressed setose, without scale-like appendages, setae up to 1·5 mm. long, fig. 12/6 (**T4**) . . . . . 30. *D. alata*

Stem terete or angled but not winged:

Decumbent herbs; leaves ovate-suborbicular to ovate-lanceolate; flowers usually solitary or in 2–4-flowered cymes:

Apex of calyx-lobes setose, without appendages; calyx-tube of young flowers with simple hairs, sometimes the hairs bulbous-based or arising from a short appendage (fig. 8/5) . . . . . 5. *D. decumbens*

Apex of calyx-lobes with a linear setose appendage; calyx-tube of young flowers with simple and branched hairs and with ligulate appendages bearing an apical tuft of hairs, the appendages ± as long as the hairs, sometimes very short, caducous and leaving a scar that is usually visible on older flowers (fig. 8/6) . . . . . . 6. *D. rotundifolia*

Erect herbs, shrubs or small trees:

Shrubs or small trees with persistent oblong-ovate calyx-lobes less than 6 mm. long:

Calyx glabrous apart from margins of lobes . . . . . . 9 *D. caloneura*

Calyx hairy:

Calyx-tube clothed in setose appendages, sometimes intermixed with simple setae, setae 2–3 mm. long (fig. 9/3) . . . . . . 13. *D. aprica*

* Intermediate forms with indumentum mainly appressed but with some spreading hairs may represent *D. debilis* × *D. phaeotricha*.

Calyx-tube not as above:
Calyx-tube 8–9 mm. long, with small simple setae mixed with setose appendages and sometimes with glandular setae (fig. 9/4, 5); leaf-lamina elliptic-lanceolate to elliptic, base truncate to subcordate . 12. *D. melleri*
Calyx-tube 4–7 mm. long; leaf-lamina oblong-ovate, ovate or elliptic, base cuneate to subcordate:
Calyx-tube campanulate, 4–4·5 mm. long, 4 mm. in diameter, sparsely setose; pedicels 3 mm. long . . . . . . 15. *D. dichaetantheroides*
Calyx-tube cylindric-campanulate, 5–7 mm. long, 4·5–5 mm. in diameter (fig. 9/9); pedicels 5–8 mm. long . . . 16. *D. polyantha*
Shrubs or herbs, sometimes woody; calyx-lobes usually more than half as long as the tube, if shorter then lobes triangular:
Calyx-lobes (figs. 8/7, 9/1) triangular-ligulate to triangular-lanceolate, 3–5 mm. long; intersepalar appendages absent:
Calyx-lobes persistent, 4–5 mm. long; plant canescent . . . . 7. *D. canescens*
Calyx-lobes caducous, 3 mm. long, with caducous, stalked setose peltate scales mixed with single or compound setae . . . . . 8. *D. brazzae*
Calyx-lobes not as above:
Calyx glabrous, sepals reflexed, longer than the tube; an erect herb with solitary flowers . . . . 4. *D. seretii*
Calyx variously ornamented with hairs and appendages:
Calyx-tube covered by swollen reflexed scale-like appendages bearing 1 or several very short setae, or scales obscured by long hairs (fig. 11/2–4) . . . 22. *D. pachytricha*
Calyx-tube not as above:
Apex of calyx-lobes long-setose (fig. 10/1–3); calyx-tube ± cylindrical, with ligulate appendages, setose at the apex; neck of tube distinctly elongating in fruit; flowers densely clustered:
Calyx-tube 6 mm. long, 4 mm. in diameter in flower; fruiting calyx 7–10 mm. long, 4·5 mm. in diameter . . 17. *D. senegambiensis*

Calyx-tube 8–10 mm. long, 5·5 mm. in diameter in flower; fruiting calyx 7–11 mm. long, 6 mm. in diameter . 18. *D. densiflora*

Apex of calyx-lobes not long-setose (calyx-lobes not known in *D. sessili-cordata*); neck of fruiting calyx not elongating and flowers not densely clustered:

Indumentum of calyx-tube densely silvery-sericeous, generally ± obscuring the scale-like appendages when present (figs. 11/1, 5, 6 & 12/1–3); leaves usually densely silvery-sericeous beneath:

Leaf-lamina sessile, cordate at base; flowers solitary . 21. *D. sessili-cordata*

Leaf-lamina not sessile and cordate at the base; flowers few or solitary at the apex of branches; bracts foliaceous or scarious, deciduous after anthesis:

Leaves conduplicate and reflexed, appressed silvery pilose or sometimes with cobweb-like indumentum on both surfaces . . 23. *D. simonis-jamesii*

Leaves not as above:

Petiole 1–6 cm. long; lamina ovate to ovate-lanceolate, 4–24 cm. long, 2·5–10·5 cm. wide, bullate and densely long-setose above, densely villous beneath . 24. *D. trothae*

Petiole up to 1 cm. long; lamina not bullate, nor long-setose:

Calyx-tube 7–10 mm. long at flowering; flowers subsessile:

Calyx-tube 7–8 × 4·5–5·5 mm., subglobose, densely appressed-sericeous, almost concealing the small scale-like appendages to which the hairs are attached (fig. 12/1); flowers 1–3, cymose, surrounded by reduced leaves . 25. *D. cryptantha*

Calyx-tube 9–10 × 5–6 mm., densely covered with oblong to obtriangular setose scales (fig. 12/2); flowers in terminal paniculate cymes; bracts caducous . . 26. *D. perkinsiae*

Calyx-tube 12–14 × 8–12 mm., densely appressed sericeous-villous, ± concealing the scale-like appendages (fig. 12/3); flowers solitary, terminal, ± concealed at first by the leaves; bracts often reddish, tardily deciduous . . 27. *D. speciosa**

Indumentum not as above; calyx-tube with ligulate or stalked capitate appendages bearing silky bristles or setae (fig. 10/4–6); inflorescences paniculate:

Calyx-tube 7–13 mm. long, 4–6 mm. in diameter, with short bristles intermixed with short, broad ligulate appendages bearing silky bristles, or appendages reduced to a cushion; leaf-lamina lanceolate to oblong-lanceolate, 4–16 cm. long, inconspicuously rugulose, sparsely and softly setulose or densely appressed pubescent above . . 19. *D. princeps*

Calyx-tube 8–10 mm. long, 6–7 mm. in diameter,

* *D. ruandensis* Engl. from Rwanda, Burundi and Zaire occurs along the border with Rwanda and Burundi. The rather brief original description is amplified by A. & R. Fernandes in Bol Soc. Brot., sér. 2, 34: 61, t. 3 (1960). *D. ruandensis* differs from *D. speciosa* by the rather larger leaves and calyx (lobes ± 17 mm. long).

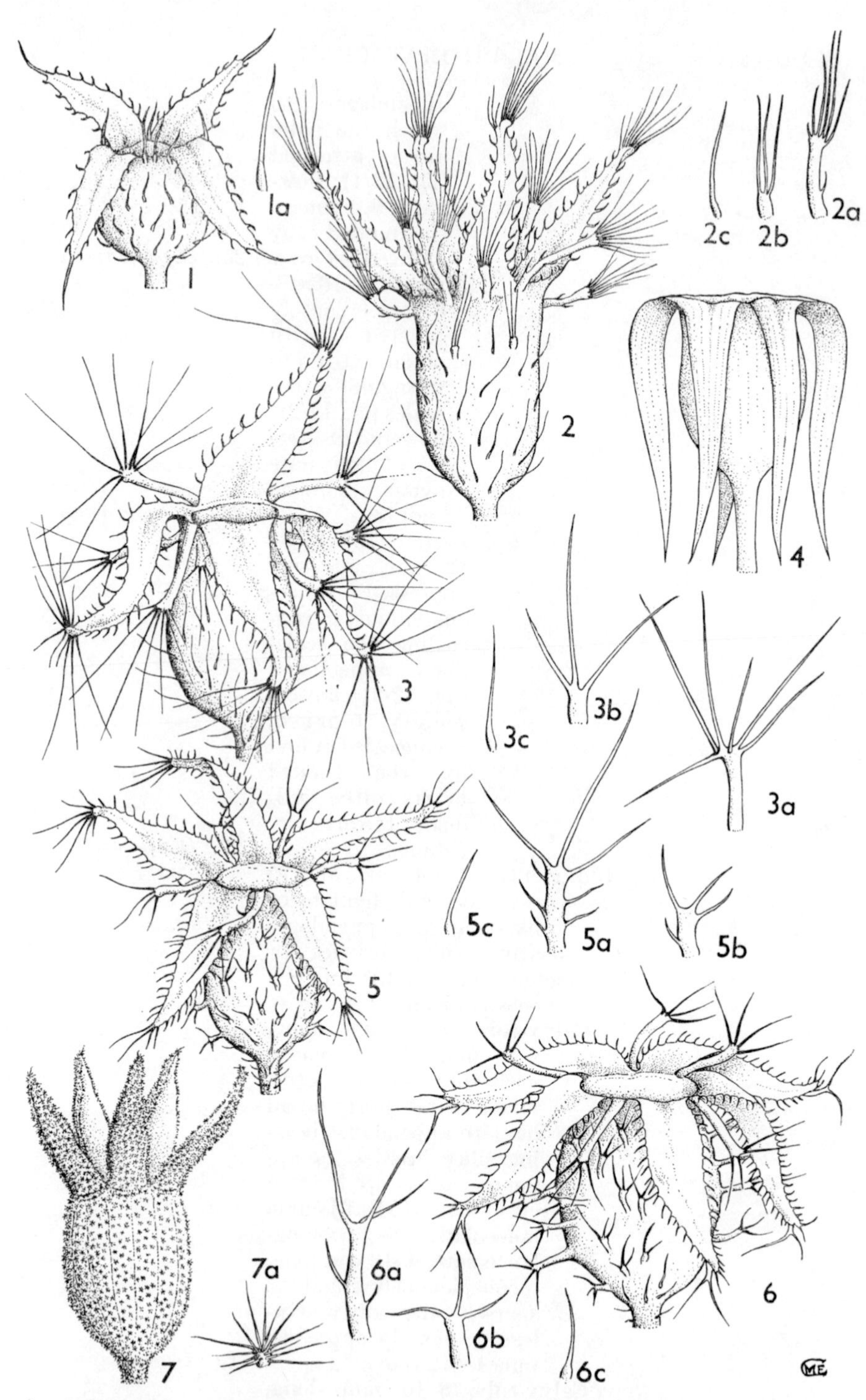

FIG. 8. Calyx and appendages from species of *Dissotis* subgenera *Osbeckiella*, *Heterotis* and *Argyrella*. **1,** *D. tisserantii*, × 8 (1a, appendage from calyx-lobe, × 16); **2,** *D. debilis* var. *debilis*, × 6 (2a, appendage from opposite and below calyx-lobe, × 8; 2b, from upper part of calyx-lobe, × 8; 2c, from lower part of calyx-tube, × 8); **3,** *D. phaeotricha* var. *phaeotricha*, × 6 (3a, appendage from opposite and below calyx-lobe, × 8; 3b, from upper part of calyx-tube, × 8; 3c, from lower part of calyx-tube. × 8); **4,** *D. seretii* var. *gracilifolia*, × 3; **5,** *D. decumbens*, × 4 (5a, appendage from opposite and below calyx-lobe, × 8; 5b, from upper and lower part of calyx-tube, × 8; 5c, from base of calyx-tube, × 8); **6,** *D. rotundifolia*, × 4 (6a, appendage from opposite and below calyx-lobe, × 8; 6b, from upper and lower part of calyx-tube; 6c, from base of calyx-tube, × 8); **7,** *D. canescens*, × 4 (7a, appendage from calyx-tube, × 8). 1, from *A. S. Thomas* 4071; 2, from *Milne-Redhead & Taylor* 10704; 3, from *Richards* 19177; 4, from *Azuma* 1014; 5, from *Mahon*; 6, from *Lye* 254; 7, from *Drummond & Hemsley* 4511. Drawn by Mrs. M. E. Church.

densely covered with 2·5 mm. long capitate appendages bearing setae 2·5 mm. long; leaf-lamina ovate-lanceolate, 2·5–11·5 cm. long, bullate and densely setose above . 20. *D. denticulata*

SYNOPSIS OF SUBGENERA AND SECTIONS

A. Flowers 4-merous; calyx-lobes persistent or tardily deciduous.

Subgen. 1. **Osbeckiella** *A. & R. Fernandes* in Bol. Soc. Brot., sér. 2, 43: 285 (1969). Annual herbs or perennial suffrutices. Inflorescence usually ± dense and capituliform; bracts small, persistent. Calyx-tube (fig. 8/1–3) cylindric-campanulate, ± densely clothed with simple hairs only or intermixed with appendages that are penicillate-setose at the apex; lobes persistent or tardily deciduous, alternating with teeth that are multi-setose at the apex or with simple long-subulate teeth. Stamens 8, markedly unequal (except in "osbeckioid" forms); 4 large stamens with arcuate connectives, each with a long anterior appendage that is entire or emarginate or strongly bilobed at the apex; 4 short stamens with short connectives, each with small bilobed appendages. Seeds ± 0·5 mm. in diameter. Species 1–3.

A. Flowers 5-merous, rarely 4-merous, then calyx-lobes caducous and connective appendage not as above.
 B. Calyx-lobes persistent; stamens with connective-appendage clavate or bluntly lobed.
  C. Flowers solitary or in few-flowered cymes; bracts caducous.

Subgen. 2. **Heterotis** (*Benth.*) *A. & R. Fernandes* in Bol. Soc. Brot., sér. 2, 43: 286 (1969). Perennial herbs or subshrubs. Flowers 5-merous (also 4-merous but not in Flora area). Calyx-tube (fig. 8/4–6) ovoid or oblong, glabrous or with simple hairs or ± sparse appendages (which have branched, stellate hairs at the apex) or sometimes only 5 ciliate subsepalar appendages; lobes persistent, membranous, reflexed apex with 1–many bristles. Stamens 8 (but not in the Flora area) or 10, markedly unequal (except in "osbeckioid" forms); anthers linear-falcate, rostrate; large stamens with elongated connectives, each with an entire or deeply lobed anterior appendage (the lobes broad and obtuse) and also a posterior bituberculate appendage; short stamens with shorter connectives and similar but smaller appendages. Ovary 4–5-locular. Seeds 1–1·25 mm. in diameter. Species 4–6.

  C. Flowers in panicles.
   D. Woody herbs; usually ± stellate-pubescent.

Subgen. 3. **Argyrella** (*Naud.*) *A. & R. Fernandes* in Bol. Soc. Brot., sér. 2, 43: 287 (1969). Erect branched herbs, shortly stellate-pubescent with stellate hairs only or mixed with simple often glandular hairs. Inflorescence of short few-flowered panicles or of larger leafy many-flowered terminal panicles. Calyx-tube (fig. 8/7) campanulate; lobes ± ovate-acute, ± equalling the tube, often with simple subulate teeth, not stellate-setose at the apex; intersepalar appendages absent or reduced. Stamens 10, unequal; anthers linear-subulate, recurved, undulating on the anterior side; 5 large stamens with elongated connectives which have a clavate appendage which is either 3-angled, truncate or emarginate; 5 short stamens with very short connectives and shortly bilobed appendage. Ovary 5-locular, the lower half adherent to the calyx-tube by septa, apex free, 5-lobed, tomentose. Seeds ± 0·75 mm. in diameter. Species 7.

   D. Trees or shrubs.

Subgen. 5.* **Dissotidendron** *A. & R. Fernandes* in Bol. Soc. Brot., sér. 2, 43: 289 (1969). Inflorescence of few-flowered apical cymes forming ± large panicles; flowers 5-merous; bracts caducous. Calyx-tube (fig. 9/2–9) cylindrical-campanulate, glabrous or with simple setae, or scale-like appendages with glandular or woolly tomentum; lobes persistent; intersepalar appendages generally absent, rarely present. Stamens 10, markedly unequal; 5 large stamens with long arcuate connectives; each with a bilobed anterior appendage (the lobes broad and obtuse); 5 small stamens with shorter

* Subgenus 4 comes under dichotomy E.

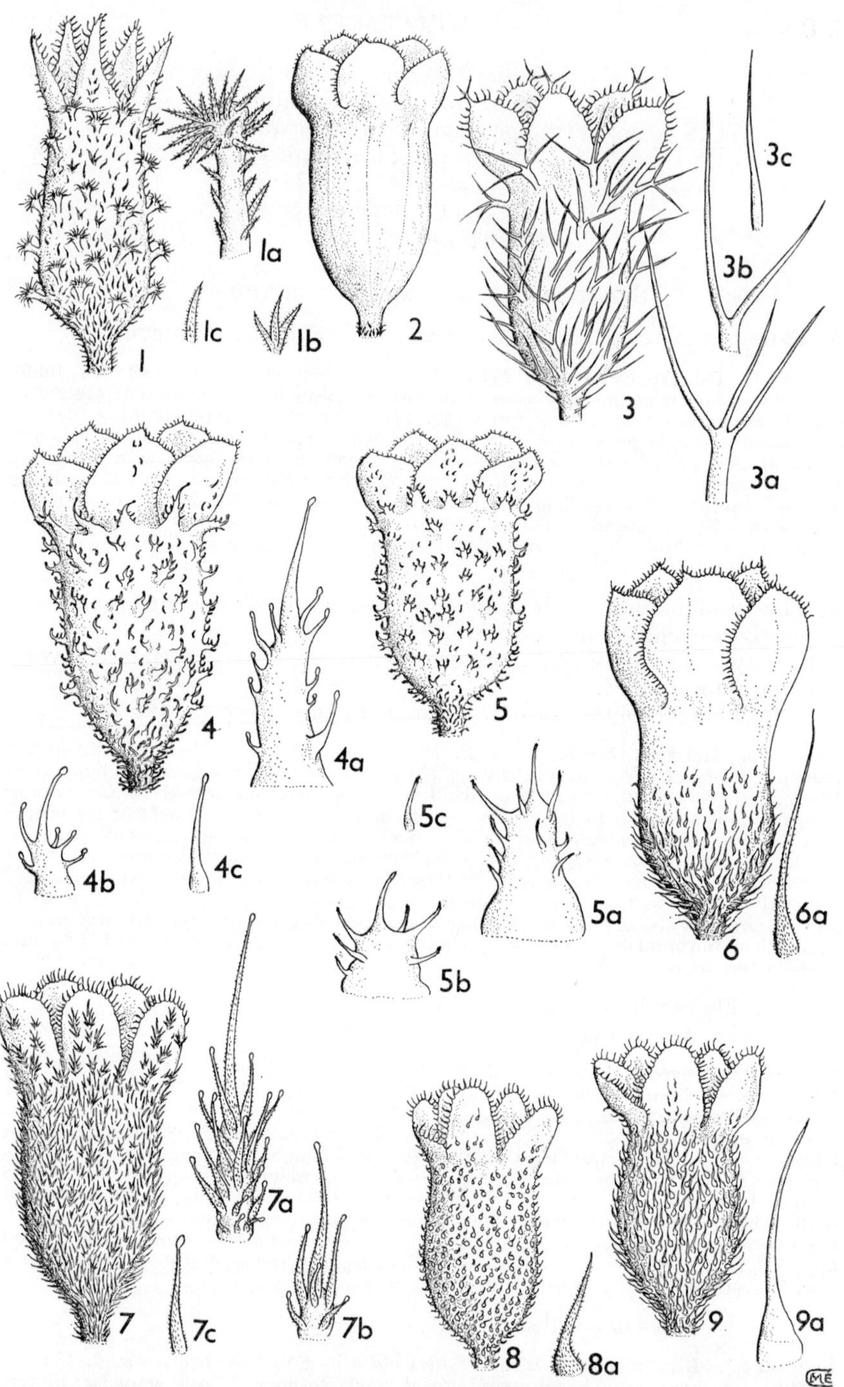

FIG. 9. Calyx and appendages from species of *Dissotis* subgenera *Dupineta* and *Dissotidendron*. **1,** *D. brazzae*, × 4 (1a-c, appendages from calyx-tube, × 16); **2,** *D. caloneura* var. *caloneura*, × 3; **3,** *D. aprica*, × 3 (3a, intersepalar appendage, × 16; 3b, c, appendages from calyx-tube, × 16); **4,** *D. melleri* var. *greenwayi*, × 3 (4a, intersepalar appendage, × 16; 4b, c, appendages from calyx-tube, × 16); **5,** *D. melleri* var. *melleri*, × 3 (5a, intersepalar appendage, × 16; 5b, c, appendages from calyx-tube, × 16); **6,** *D. arborescens*, × 3 (6a, appendage from calyx-tube, × 16); **7,** *D. glandulicalyx*, × 3 (7a, appendage from calyx-lobe and top of tube, × 16; 7b, c, appendages from calyx-tube, × 16); **8,** *D. bussei*, × 3 (8a, appendage from calyx-tube, × 16); **9,** *D. polyantha*, × 3 (9a, appendage from calyx-tube, × 16). 1, from *Drummond & Hemsley* 4780; 2, from *Richards* 12796; 3, from *Semsei* 696; 4, from *Richards* 6464; 5, from *Kerfoot* 4249; 6, from *Mathias & Taylor* A. 92; 7, from *Harley* 9352; 8, from *Bally* 7899; **9,** from *Gibbon & Pócs* 6052. Drawn by Mrs. M. E. Church.

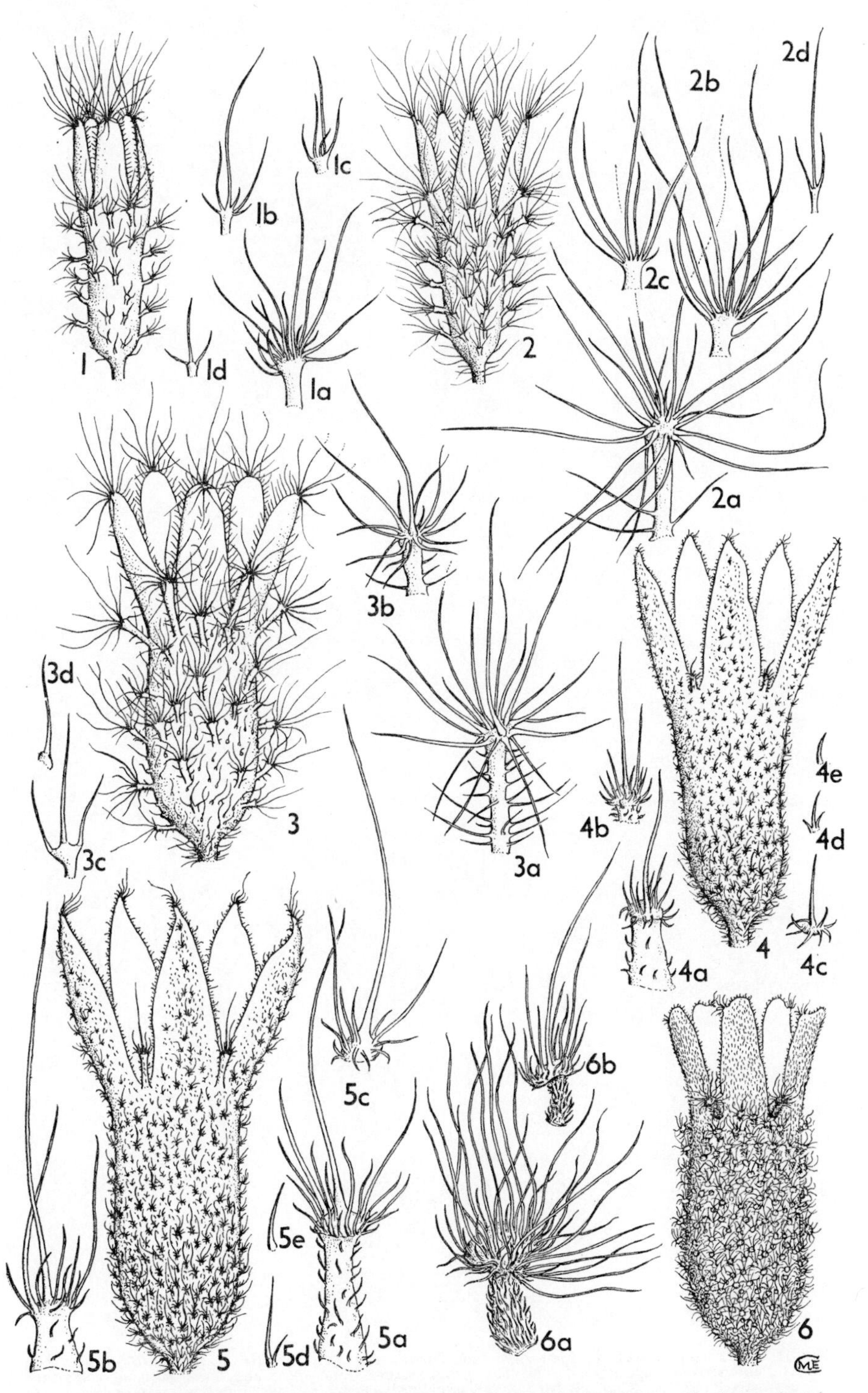

FIG. 10. Calyx and appendages from species of *Dissotis* subgenus *Dissotis*. **1,** *D. senegambiensis* var. *alpestris*, × 3 (1a, intersepalar appendage, × 6; 1b–d, other appendages, × 6); **2,** *D. senegambiensis* var. *senegambiensis*, × 3 (2a, intersepalar appendage, × 6; 2b–d, other appendages, × 6); **3,** *D. densiflora*, × 3 (3a, intersepalar appendage, × 6; 3b–d, other appendages, × 6); **4,** *D. princeps* var. *candolleana*, × 3 (4a, intersepalar appendage, × 12; 4b–e, other appendages, × 24); **5,** *D. princeps* var. *princeps*, × 3 (5a, intersepalar appendage, × 12; 5b–e, other appendages, × 24); **6,** *D. denticulata*, × 3 (6a, intersepalar appendage, × 12; 6b, appendage from calyx-tube, × 12). 1, from *Drummond & Hemsley* 1501; 2, from *Renvoize & Abdallah* 1915; 3, from *Richards* 8819; 4, from *Stolz* 237; 5, from *Milne-Redhead & Taylor* 9009; 6, from *Greenway* 8429. Drawn by Mrs. M. E. Church.

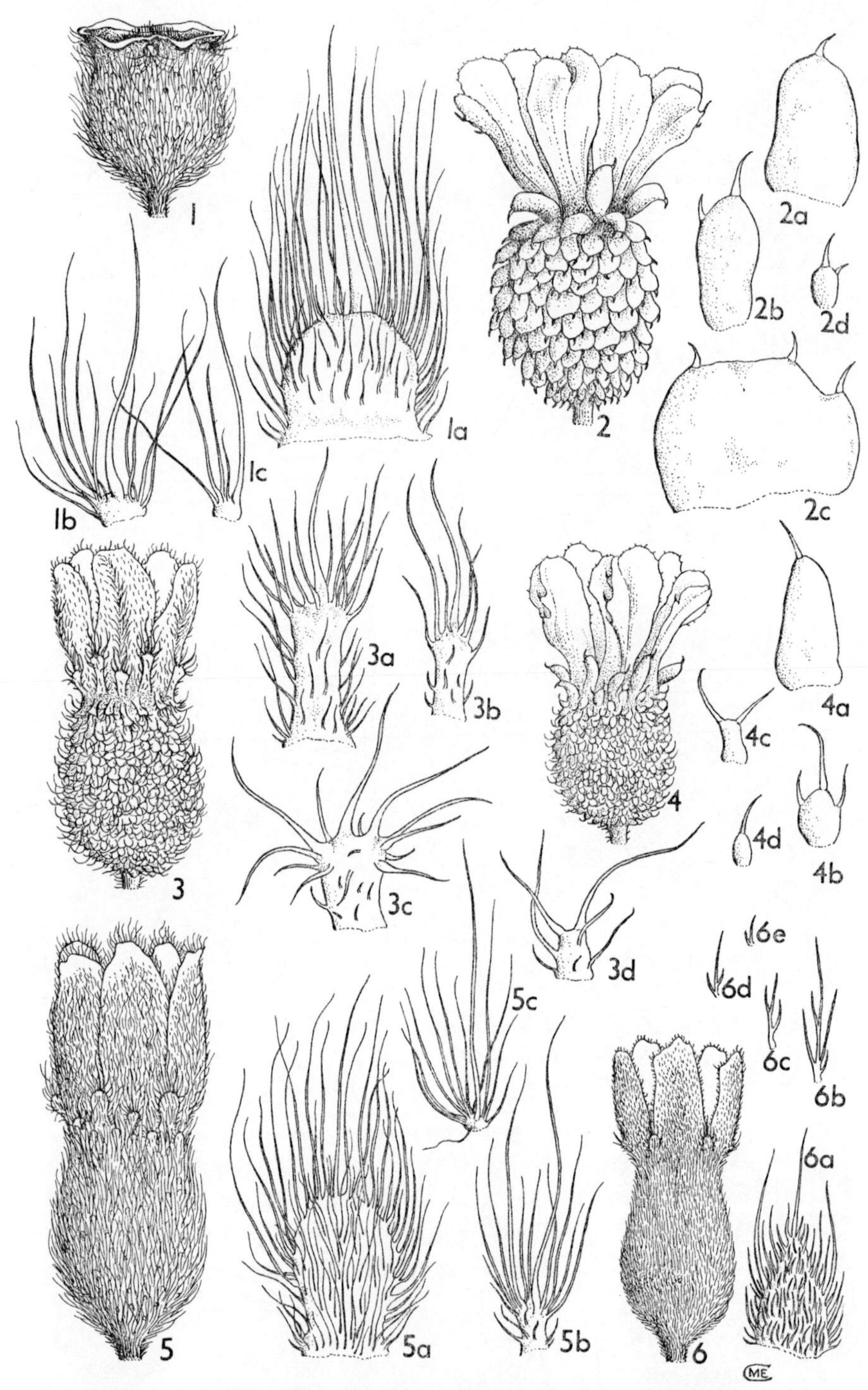

FIG. 11. Calyx and appendages from species of *Dissotis* subgenus *Dissotis*. **1,** *D. sessili-cordata*, × 2 (1a, intersepalar appendage, × 6; 1b, c, appendages from upper and lower parts of calyx-tube respectively, × 8); **2,** *D. pachytricha* var. *grandisquamulosa*, × 2 (2a, intersepalar appendage, × 6; 2b–d, appendages from upper, middle and lower parts of calyx-tube respectively, × 8); **3,** *D. pachytricha* var. *orientalis*, × 2 (3a, intersepalar appendage, × 6; 3b–d, appendages from upper, middle and lower parts of calyx-tube respectively, × 8); **4,** *D. pachytricha* var. *pachytricha*, × 2 (4a, intersepalar appendage, × 6; 4b–d, appendages from upper, middle and lower parts of calyx-tube respectively, × 8); **5,** *D. simonis-jamesii*, × 2 (5a, intersepalar appendage, × 2; 5b, appendage from upper half of calyx-tube, × 8; 5c, appendage from lower half of calyx-tube, × 8); **6,** *D. trothae*, × 3 (6a, intersepalar appendage, × 12; 6b–e, appendages from successively lower parts of calyx-tube, × 8). 1, from *Semsei* 111; 2, from *Bullock* 3283; 3, from *Milne-Redhead & Taylor* 10740; 4, from *Newbould & Harley* 4549; 5, from *Bullock* 3740; 6, from *Richards* 10180. Drawn by Mrs. M. E. Church.

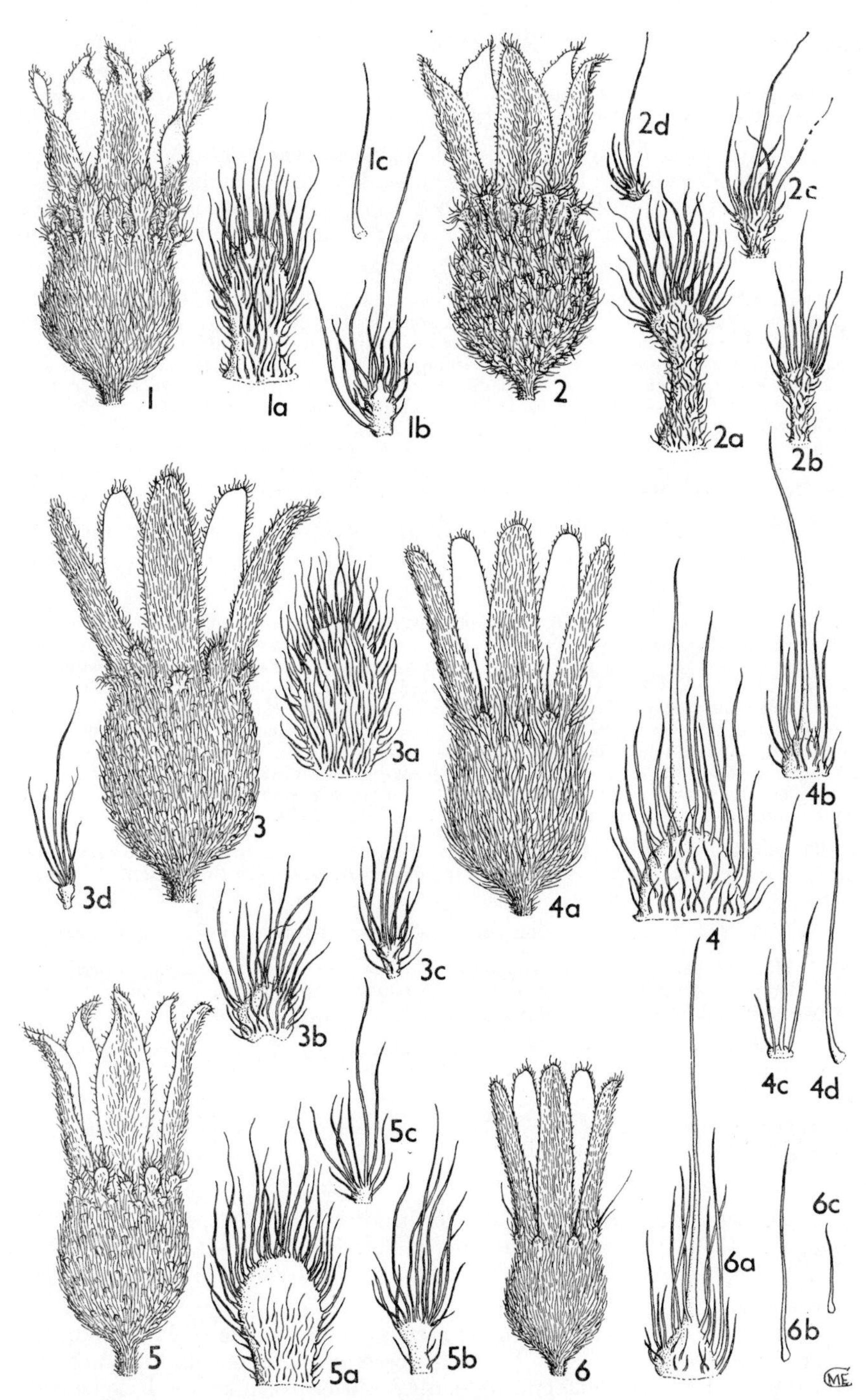

FIG. 12. Calyx and appendages from species of *Dissotis* subgenus *Dissotis*. **1,** *D. cryptantha*, × 3 (1a, intersepalar appendage, × 6; 1b, appendage from upper part of calyx-tube, × 8; 1c, appendage from lower on calyx-tube, × 8); **2,** *D. perkinsiae*, × 2 (2a, intersepalar appendage, × 6; 2b–d, appendages from upper, lower and basal parts of calyx-tube respectively, × 8); **3,** *D. speciosa*, × 2 (3a, intersepalar appendage, × 6; 3b–d, appendages from upper, lower and basal parts of calyx-tube respectively, × 8); **4,** *D. pterocaulos*, × 2 (4a, intersepalar appendage, × 6; 4b, appendage from upper part of calyx-tube, × 8; 4c, d, appendages from lower on calyx-tube, × 8); **5,** *D. formosa*, × 2 (5a, intersepalar appendage, × 6; 5b, appendage from calyx-tube, × 8; 5c, appendage from base of calyx-tube, × 8); **6,** *D. alata*, × 3 (6a, intersepalar appendage, × 12; 6b, c, appendages from calyx-tube, × 8). 1, from *Milne-Redhead & Taylor* 10729; 2, from *Dawkins* 451; 3, from *Haarer* 2318; 4, from *Verdcourt* 3427; 5, from *Milne-Redhead & Taylor* 9924A; 6, from *Verdcourt & Oteke* 2861. Drawn by Mrs. M. E. Church.

connectives and smaller appendages. Ovary 5-locular, ovoid, rarely free, apex setose. Seeds ± 0.5 mm. in diameter. Species 9–16.

B. Calyx-lobes caducous, or if sometimes subpersistent, then lobes of the appendage of the large stamens attenuate and ± acute.

E. Fruiting calyx with a long neck and distinct longitudinal ribs; stamens with bilobed clavate appendages; intersepalar appendages absent.

Subgen. 4. **Dupineta** (*Raf.*) *A. & R. Fernandes* in Bol. Soc. Brot., sér. 2, 43: 288 (1969). Perennial herbs. Inflorescence of paniculate cymes; flowers 5-merous. Calyx-tube (fig. 9/1) ovoid or urceolate, with ± dense simple spinulose hairs only or intermixed with caducous clawed scale-like appendages with compound stellate-setose hairs; lobes caducous; intersepalar appendages absent. Stamens 10, markedly unequal (subequal in " osbeckioid " forms); 5 large stamens with long connectives, each with a thick bilobed anterior appendage; 5 small stamens with short connectives and a bilobed anterior appendage. Ovary 5-locular, adhering to the calyx-tube by 5 septa, apex with long simple setose hairs; fruiting calyx developing a long neck with distinct longitudinal ribs. Seeds ± 0·5 mm. in diameter. Species 8.

E. Fruiting calyx not as above; stamen-appendage bifid with attenuate ± acute segments; intersepalar appendages usually present.

Subgen. 6. **Dissotis**; *A. & R. Fernandes* in Bol. Soc. Brot., sér. 2, 43: 289 (1969). Perennial herbaceous or woody herbs or shrubs. Inflorescence of paniculate cymes of flowers solitary at the apex of branches; flowers (4–)5(–6)-merous; bracts caducous. Calyx-tube (figs. 10–12) ovoid to cylindrical, glabrous or ± densely covered with simple setae or scales or appendages; lobes caducous, sometimes tardily so, apex often multi-setose; intersepalar appendages usually present. Stamens markedly unequal (subequal in " osbeckioid " forms); large stamens with long arcuate connectives, each with an anterior appendage, often bifid with segments that are attenuated and ± acute; short stamens with short connectives and similar but smaller bifid or bituberculate appendages. Seeds ± 0·5 mm. in diameter. Species 17–30.

The subgenus is divided further by A. & R. Fernandes (*loc. cit.*) into rather ill-defined sections as follows, but this scheme is not strictly adhered to in the sequence of species adopted here.

Sect. **Dissotis.** Calyx-tube ± long and densely setose, with pedicellate appendages.

Sect. **Macrocarpae** *A. & R. Fernandes.* Flowers few or solitary at the apex of the branches; bracts foliaceous or scarious, deciduous after anthesis. Indumentum of calyx-tube ± densely silvery-sericeous.

Sect. **Sessilifoliae** *A. & R. Fernandes.* Calyx-tube glabrous or sparsely hairy or with appendages bearing 1(–3) apical setae. Leaves parchment-like, often sessile, glabrous or sparsely hairy.

Sect. **Squamulosae** *A. & R. Fernandes.* Calyx-tube clothed with sessile scales, these small, pectinate-ciliate or multifid.

1. **D. tisserantii** *Jacques-Félix* in Bull. Mus. Hist. Nat., Paris, sér. 2, 10: 633, fig. 3 (1938); A. & R. Fernandes in Bol. Soc. Brot., sér. 2, 34: 59 (1960). Type: Central African Republic, Bambari, *Tisserant* 1256 (P, holo., BM, iso.!)

Annual herb (3–)6–20 cm. high; stem slender, simple or branched, very sparsely appressed setose. Leaves subsessile; lamina elliptic-lanceolate, 0·5–0·8 cm. long, 0·2–0·25 cm. wide, apex obtuse to subacute, base cuneate, margin entire or serrulate, glabrous; midrib scarcely visible above, subprominent beneath; midrib and 1 pair of basal longitudinal nerves subprominent beneath. Inflorescence a 2–5-flowered terminal cluster; pedicel 1–2 mm. long; bracts linear-lanceolate. Calyx-tube (fig. 8/1, p. 28) campanulate, 2 mm. long, 2 mm. in diameter, sparsely pilose; lobes narrowly triangular, 2·5 mm. long, margin serrulate-ciliate; intersepalar appendages bristle-like. Petals pink, obovate, ± 5 mm. long, 4 mm. wide. Stamens 8, unequal; 4 large stamens with anthers ± 1 mm. long, free part of connective arcuate ± 1·8 mm. long, the solitary anterior appendage 1·8 mm. long, bilobed at the apex, filament 2·5 mm. long; 4 short stamens with anthers ± 1 mm. long, connective ± 5 mm. long, anterior appendage 0·4 mm. long, emarginate at the apex, filament 2·5 mm. long. Ovary 4-locular, subglobose, lower half adherent to the calyx-tube, apex with scattered bristles. Seeds 0·5 mm. long, 0·3 mm. in diameter, minutely tuberculate, the tubercles in curved rows.

UGANDA. W. Nile District: Jebel Midigo, 26 Nov. 1941, *A. S. Thomas* 4071!
DISTR. **U**1; Zaire and Central African Republic
HAB. Seepage zone among granite rocks; 1050 m.

NOTE. Superficially this plant may be mistaken for *Antherotoma naudinii* Hook. f., from which it may be readily distinguished by the shape of the anthers.

2. **D. debilis** (*Sond.*) *Triana* in Trans. Linn. Soc. 28: 58, t. 4/44a, b (1872); Cogn. in A. & C. DC., Monogr. Phan. 7: 367 (1891); Gilg in E.M. 2: 14, t. 2/D (1898); V.E. 3 (2): 748 (1921); F.P.S. 1: 192 (1950); A. & R. Fernandes in Mem. Soc. Brot. 11: 15, 70 (1956) & in Bol. Soc. Brot., sér. 2, 34: 182 (1960). Type: South Africa, Transvaal, Magalisberg, *Zeyher* 539 (S, holo., BM, K, iso.!)

Erect woody herb from a perennial rootstock or annual (semiprostrate and prostrate forms are known to the south of the Flora area), 9–55 cm. high; stem simple or branched, appressed pilose. Leaf-lamina ovate, oblong-ovate, ovate-lanceolate or linear-lanceolate, 1–6 cm. long, 0·2–1·6 cm. wide, apex acute to obtuse, base shortly attenuate to cuneate, margin entire to subserrate, bristles ± appressed to the margin, appressed setose on both surfaces; midrib and 1 pair of basal longitudinal nerves lightly impressed above, subprominent beneath; petiole 0–4 mm. long. Inflorescence terminal, capitate, many-flowered with 4–6 persistent bracteose foliage leaves; flowers 4-merous, shortly pedicellate. Calyx-tube (fig. 8/2, p. 28) cylindric-campanulate, 3·5–5 mm. long, 2–3 mm. in diameter, lightly setose; lobes subulate, 3–4 mm. long, margin ciliate, otherwise glabrescent except for a tuft of setae at the apex; intersepalar appendages linear, 2 mm. long, with stellately arranged setae at the apex. Petals mauve, broadly obovate, 8–14 mm. long, 8–13 mm. wide, slightly emarginate. Stamens 8, markedly unequal (except in forma *osbeckioides*); 4 large stamens with anthers 4·5 mm. long, arcuate connective 5–7 mm. long, bilobed appendage ± 1·5 mm. long, filament 7 mm. long; 4 small stamens with anthers 3 mm. long, connective ± 0·3 mm. long, appendage ± 0·2 mm. long, filament 3 mm. long; forma *osbeckioides* with anthers 4·5 mm. long, connective 0·7 mm. long, appendage 0·2 mm. long, filament 7 mm. long. Ovary with an apical crown of setae.

KEY TO INFRASPECIFIC VARIANTS

Plants with stems arising from a perennial rootstock:
  Leaves ovate, oblong-ovate or ovate-lanceolate, 1–6 cm. long, 0·8–1·6 cm. wide, base shortly attenuate . . . . var. **debilis**
  Leaves linear-lanceolate, 1·5–5 cm. long, 0·2–0·5 cm. wide, base cuneate . . . . . var. **lanceolata**
Plants annual . . . . var. **postpluvialis**

var. **debilis**; A. & R. Fernandes in An. Junta Invest. Ultramar 10: 17 (1955) & in Mem. Soc. Brot. 11: 16, 71 (1956) & in Bol. Soc. Brot., sér. 2, 34: 182 (1960) & in C.F.A. 4: 137 (1970)

Erect herb 9–40 cm. high. Leaf-lamina ovate, oblong-ovate or ovate-lanceolate, 1–6 cm. long, 0·8–1·6 cm. wide, base shortly attenuate. Stamens subequal or markedly unequal.

KENYA. Uasin Gishu District: Kipkarren R., *Brodhurst-Hill* 367 !; Kwale District: Vanga, Majoreni, Nov. 1929, *R. M. Graham* LL.731 in *F.D.* 2183 !
TANZANIA. Ufipa District: 48 km. Mbala [Abercorn] to Sumbawanga, 3 Mar. 1951, *Bullock* 3739 !; Dodoma District: Rungwa Game Reserve, Sulangi, 21 Jan. 1969, *Ismail* 4236 !; Tunduru District: Litungura, 6 June 1956, *Milne-Redhead & Taylor* 10669 !
DISTR. **K**3, 7; **T**4–8; Zaire, Mozambique, Malawi, Zambia, Rhodesia, Botswana, Angola and South Africa
HAB. Valley grassland; 10–1800 m.

SYN. *Osbeckia debilis* Sond. in Linnaea 23: 47 (1850), *non* Naud. (1850)
O. phaeotricha Hochst. var. *debilis* (Sond.) Sond. in Fl. Cap. 2: 519 (1862)
Dissotis penicillata Gilg in E.M. 2: 14 (1898); V.E. 3 (2): 748 (1921). Type: Angola, Huila, *Antunes* 126 (B, holo. †, COI, lecto.)
D. paludosa Engl. in V.E. 3 (2): 748 (1921), *nom. nud.*

NOTE. A. & R. Fernandes, in An. Junta Invest. Ultramar 10: 17, 18 (1955) & in Bol. Soc. Brot., sér. 2, 34; 182 (1960), recognize forma *debilis* with unequal stamens and forma *osbeckioides* A. & R. Fernandes with subequal stamens. Both forms occur in the Flora area and have been found as mixed populations (see *Milne-Redhead & Taylor* 10479 & 10480). Plants from **T**6, such as *Haerdi* 328/0 from Ulanga District, Ifakara, with an indumentum intermediate between appressed and spreading, have been regarded as a hybrid between var. *debilis* and *D. phaeotricha* (Hochst.) Hook. f.

var. **lanceolata** (*Cogn.*) *A. & R. Fernandes* in Bol. Soc. Brot., sér. 2, 29: 49, t. 3 (1955) & in Mem. Soc. Brot. 11: 19 (1956) & in Bol. Soc. Brot., sér. 2, 34: 183 (1960) & in C.F.A. 4: 139 (1970). Type: Angola, Pungo Andongo, *Welwitsch* 910 (G, holo., BM !, BR, COI, K !, LISU, P, iso.)

Erect herb 19–55 cm. high. Leaf-lamina linear-lanceolate, 1·5–5 cm. long, 0·2–0·5 cm. wide, base cuneate. Stamens subequal or markedly unequal.

TANZANIA. Ufipa District: Lake Kwela, 11 Mar. 1959, *Richards* 11140 !; Mbeya District: Mbozi, May 1935, *Horsbrugh-Porter* !; Masasi, 22 Apr. 1935, *Schlieben* 6360 !
DISTR. **T**4, 7, 8; Zaire, Malawi, Zambia, Rhodesia, Angola and South Africa
HAB. Valley grassland; 760–1650 m.

SYN. *D. lanceolata* Cogn. in A. & C. DC., Monogr. Phan. 7: 366 (1891)

NOTE. A. & R. Fernandes, in Bol. Soc. Brot., sér. 2, 29: 49 (1955) & in 34: 183 (1960), recognize forma *lanceolata* with unequal stamens and in *op. cit.* 29: 50 (1955), forma *subisandra* A. & R. Fernandes with subequal stamens; so far only forma *lanceolata* has been recorded in the Flora area.

var. **postpluvialis** (*Gilg*) *A. & R. Fernandes* in Garcia de Orta 2: 171 (1954) & in Bol. Soc. Brot., sér. 2, 34: 183 (1960). Type: Sudan, Bahr et Ghazal Province, Ghattas's Zeriba, Nov. 1870, *Schweinfurth* 4289 (B, holo. †, K, iso. !)

Annual plant, 10–23 cm. high; stem simple or scarcely branched. Leaf-lamina oblong-lanceolate, 2·5–5 cm. long, 1–1·6 cm. wide, base shortly attenuate. Stamens subequal or markedly unequal.

UGANDA. W. Nile District: Mt. Otzi, Oct. 1951, *Chenery* 7 ! & Jebel Midigo, *Uganda Agr. Dept.* in *Kew Spirit No.* 14466 !
DISTR. **U**1; Sudan, Mozambique and Angola
HAB. Valley grassland

SYN. *Osbeckia postpluvialis* Gilg in E.M. 2: 6 (1898); V.E. 3 (2): 744 (1921)

NOTE. Observations are required to establish whether this variety is truly annual and not first year growth of var. *debilis*.

3. **D. phaeotricha** (*Hochst.*) *Hook. f.* in F.T.A. 2: 451 (1871)*; Triana in Trans. Linn. Soc. 28: 58 (1872); Cogn. in A. & C. DC., Monogr. Phan. 7: 367

* Earlier authors have attributed the combination to Triana, but according to Miss Raphael in Biol. J.L.S. 2: 73 (1970) the relevant volume of Trans. Linn. Soc. was published in early January 1872, whereas volume 2 of F.T.A. was published in July 1871.

(1891); Gilg in E.M. 2: 14 (1898); V.E. 3 (2): 748 (1921); F.P.S. 1: 192 (1950); F.W.T.A., ed. 2, 1: 259 (1954), pro parte excl. syn. *Osbeckia postpluvialis* Gilg; A. & R. Fernandes in An. Junta Invest. Ultramar 10: 21 (1955) & in Bol. Soc. Brot., sér. 2, 34: 183 (1960) & in C.F.A. 4: 142 (1970). Type: South Africa, Durban [Port Natal], *Krauss* 201 (B, holo. †, BM, K, iso.!)

Erect woody herb from a perennial rootstock, (12–)19–32 cm. high; stem simple or branched, densely pilose, the hairs spreading or ± appressed. Leaf-lamina oblong-ovate or oblong-lanceolate, 1·5–6·5 cm. long, 0·7–1·7 cm. wide, apex acute, base cuneate to obtuse, margin lightly crenulate, bristles ± appressed to the crenulations, ± densely setose on both surfaces; midrib and 1 pair of basal longitudinal nerves lightly impressed above, subprominent beneath; petiole 0·5–3 mm. long. Inflorescence terminal, capitate, many-flowered, subtended by 2–4 foliaceous bracts; flowers 4-merous, shortly pedicellate. Calyx-tube (fig. 8/3, p. 28) cylindric-campanulate, 3·5–4·5 mm. long, 3–3·5 mm. in diameter, densely pubescent with a mixture of very short hairs and longer bulbous-based hairs and with a few scattered linear-setose appendages near the mouth; lobes triangular-subulate, 4 mm. long, margin ciliate, otherwise glabrous except for a tuft of setae at the apex; intersepalar appendages linear, 1·5–3·5 mm. long, with a tuft of setae at the apex. Petals purple, broadly obovate, 17 mm. long, 15 mm. wide. Stamens 8, markedly unequal or subequal (forma *osbeckioides*); 4 large stamens with anthers 3·5–4·5 mm. long, free part of connective arcuate, 4·5–5·5 mm. long, bilobed appendage 2 mm. long, filament 3–3·5 mm. long; 4 small stamens with anthers 3·2–4·5 mm. long, connective produced 0·5–1·5 mm., bilobed appendage ± 0·2 mm. long, filament 3 mm. long; forma *osbeckioides* with anthers 3 mm. long, connective produced ± 0·5 mm., bilobed appendage ± 0·2 mm. long, filament 3 mm. long. Ovary with an apical crown of setae.

var. **phaeotricha**; *A. & R. Fernandes* in An. Junta Invest. Ultramar 10: 22 (1955) & in Mem. Soc. Brot. 11: 20, 73 (1956) & in Bol. Soc. Brot., sér. 2, 34: 183 (1960) & in C.F.A. 4: 142 (1970)

Uganda. Masaka District: Sese Is., Sozi, Jan. 1922, *Maitland*! & Katera, 1 Oct. 1953, *Drummond & Hemsley* 4513! & Sanje, 16 May 1971, *Lye* 6063!
Tanzania. Bukoba District: Kakindu, Oct. 1931, *Haarer* 2319!; Ulanga District: Lupiro, 8 Aug. 1955, *Anderson* 1077!; Iringa District: 32 km. Iringa–Dodoma road, 21 Sept. 1964, *Richards* 19177!
Distr. **U4**; **T1**, 4–8; widespread in tropical Africa from Senegal eastwards to the Sudan and southwards to Angola and Natal
Hab. Valley grassland; 1110–1200(–1860) m.

Syn. *Osbeckia phaeotricha* Hochst. in Flora 27: 424 (1844) & in Walp, Repert. Bot. Syst. 5: 708 (1846)
Argyrella ? phaeotricha Naud. in Ann. Sci. Nat., sér. 3, 13: 300 (1849)
[*Dissotis irvingiana* Hook. var. *irvingiana* forma *abyssinica* sensu A. & R. Fernandes in Mem. Soc. Brot. 11: 89 (1956) pro specim. *Haarer* 2319, *non* (Gilg) A. & R. Fernandes]

Note. A. & R. Fernandes in An. Junta Invest. Ultramar 10: 24 (1955) recognize forma *phaeotricha* with unequal stamens and forma *osbeckioides* A. & R. Fernandes with subequal stamens; both forms are known to occur within the same populations in Mozambique but so far forma *osbeckioides* has not been recorded within the Flora area.

var. **villosissima** *A. & R. Fernandes* in Kirkia 1: 71 (1961). Type: Zambia, Mporokoso District, Mweru, *Richards* 9065 (K, holo.!, COI, iso.)

Indumentum on stem and leaves dense, subappressed. Inflorescence ± concealed by the leaves.

Tanzania. Ufipa District: Namanyere, 4 Apr. 1950, *Bullock* 2839!; Chunya District: Chunya–Itigi road, 21 Mar. 1965, *Richards* 19792!
Distr. **T4**, 7; Zaire and Zambia
Hab. Valley grassland; 1500–1650 m.

NOTE. A. & R. Fernandes, *loc. cit.* (1961), recognize forma *villosissima* with unequal stamens and forma *osbeckioides* A. & R. Fernandes with subequal stamens; the latter so far has only been recorded from Zaire.

4. **D. seretii** *De Wild.* in Ann. Mus. Congo, Bot., sér. 5, 2: 328 (1908). Type: Zaire, Bokoyo, *Seret* 587 [581] (BR, holo.!)

Erect herb 15–20 cm. high, ± glabrous. Leaves sessile; lamina linear-lanceolate to elliptic-lanceolate, 2–7 cm. long, 0·5–1·5 cm. wide, apex obtuse or acute, base truncate, margin ciliate, glabrous above, sparsely hairy beneath; midrib impressed above, subprominent beneath. Inflorescence terminal; flowers solitary, 5-merous, pedicellate. Calyx-tube (fig. 8/4, p. 28) oblong-campanulate, 7·5 mm. long, 5 mm. in diameter, glabrous; lobes linear-lanceolate, 13–14 mm. long, reflexed; intersepalar appendages absent. Petals narrowly obovate, 22–28 mm. long, 12 mm. wide, apex shortly acuminate. Stamens 10; large stamens with anthers 7 mm. long, free part of connective arcuate, 23 mm. long, appendage 3·5 mm. long, bilobed, filament 8 mm. long; short stamens with anthers 5 mm. long, connective produced 1 mm., appendage 0·5 mm. long, filament 8 mm. long. Fruiting calyx not seen.

var. **gracilifolia** *Wickens* in K.B. 29: 141 (1974). Type: Tanzania, Kigoma District, Kabogo Mts., *Azuma* 1014 (EA, holo.!)

Leaf-lamina linear-lanceolate, 2–2·8 cm. long, 0·15–0·4 cm. wide, apex acute. Pedicels 11–13 mm. long, the flowers not concealed by the uppermost pair of leaves.

TANZANIA. Kigoma District: Kabogo Mts., Itatie, 25 Dec. 1963, *Azuma* 1014!
DISTR. **T4**; not known elsewhere (var. *seretii* occurs in Zaire)
HAB. Valley grassland; ± 1500 m.

SYN. [*D. seretii* sensu A. & R. Fernandes in Bol. Soc. Brot., sér. 2, 43: 293, t. 2 (1969), *non* De Wild. sensu stricto]

5. **D. decumbens** (*P. Beauv.*) *Triana* in Trans. Linn. Soc. 28: 58 (1872); Gilg in E.M. 2: 15 (1898); V.E. 3 (2): 754 (1921); F.P.S. 1: 194 (1950); F.W.T.A., ed. 2, 1: 257 (1954); A. & R. Fernandes in C.F.A. 4: 143 (1970). Type: Nigeria, 1786–7, *Palisot de Beauvois* (G, holo.)

Decumbent herb, rooting at the nodes, with erect branches up to 15 cm. or more high; stem lightly pubescent. Leaf-lamina ovate, broadly ovate or oblong-ovate, 1·5–3·5 cm. long, 1·2–2·7 cm. wide, apex acute, base truncate to shortly attenuate, margin slightly dentate-ciliate, sparsely pubescent on both surfaces; midrib and 1 pair of basal longitudinal nerves lightly impressed above, subprominent beneath; petiole 0·5–1·5 cm. long, pubescent. Inflorescence of a solitary flower or rarely a 2–4-flowered cyme; flowers 5-merous. Calyx-tube (fig. 8/5, p. 28) cylindrical-campanulate, 5–6 mm. long, 3·5–4 mm. in diameter, sparsely pilose; lobes linear-lanceolate, 5 mm. long, margin ciliate, setose at the apex; intersepalar appendages ligulate, 2 mm. long, setose at the apex. Petals rose-purple, obovate, 20 mm. long, 16 mm. wide. Stamens 10, unequal; large stamens with anthers 7 mm. long, free part of connective arcuate, 6–7 mm. long, bilobed appendage 2 mm. long, filament 6 mm. long; small stamens with anthers 6 mm. long, free part of connective 0·5 mm. long, bilobed appendage 0·5 mm. long, filament 5·5 mm. long.

UGANDA. Mengo District: Entebbe, 1903, *Mahon*! & Kirerema–Kipayo, Apr. 1914 (or ? 1913), *Dummer* 173!
TANZANIA. Lake Tanganyika, E. shore, Oct. 1894, *Scott Elliot* 8232!
DISTR. **U4**; **T4**; Nigeria, Cameroun, Gabon, Zaire, Sudan and Angola; introduced into Mascarene Is.
HAB. Margins of rain-forest; 1170–1200 m.

SYN. *Melastoma decumbens* P. Beauv., Fl. Owar. Benin 1: 69, t. 41 (1806)
Osbeckia decumbens (P. Beauv.) DC., Prodr. 3: 143 (1828)
Heterotis laevis Benth. in Hook., Niger Fl.: 348 (1849). Type: Nigeria, R. Nun, *Vogel* (K, holo.!)
Dissotis laevis (Benth.) Hook. f. in F.T.A. 2: 451 (1871)
D. decumbens (P. Beauv.) Triana var. *minor* Cogn. in A. & C. DC., Monogr. Phan. 7: 369 (1891). Types: Zaire, Bangala, *Hens* 164 (P, syn., K, isosyn.!) & without locality, *Thollon* (P, syn.)
D. mahonii Hook. f. in Bot. Mag. 129, t. 7896 (1903), as "*mahoni*". Type: plant grown at Kew in 1902 from seed sent from Uganda by Mahon in 1901 (K, holo.!)
D. modesta Stapf in K.B. 1906: 78 (1906). Type: plant grown at Kew in 1905 from seed sent from Uganda by Dawe (K, holo.!)

NOTE. This species can easily be confused with *D. rotundifolia*, especially the older flowering specimens. Young flowers should be examined whenever possible in order to see the indumentum.

6. **D. rotundifolia** (*Sm.*) *Triana* in Trans. Linn. Soc. 28: 58 (1872); Cogn. in A. & C. DC., Monogr. Phan. 7: 369 (1891); Gilg in E.M. 2: 15 (1898); V.E. 3 (2): 754 (1921); T.T.C.L.: 310 (1949); U.O.P.Z.: 234, fig. (1949); F.W.T.A., ed. 2, 1: 257, fig. 101 (1954); A. & R. Fernandes in Mem. Soc. Brot. 11: 23, 74 (1956) & in Bol. Soc. Brot., sér. 2, 34: 184 (1960) & in C.F.A. 4: 143 (1970). Type: Sierra Leone, *Afzelius* (BM, holo.!)

Decumbent herb, rooting at the nodes, with erect branches up to 16 cm. or more high; stem lightly to densely hirsute. Leaf-lamina ovate-suborbicular to oblong-ovate or ovate-lanceolate, 1·5–7 cm. long, 0·8–4 cm. wide, apex acute, base truncate to shortly attenuate, margin slightly crenulate-ciliate, lightly to densely pilose on both surfaces; midrib and 1(–2) pairs of basal longitudinal leaves lightly impressed above, subprominent beneath; petiole 0·5–2·5 cm. long. Inflorescence of a solitary flower or rarely 2–4-flowered cymes; flowers 5-merous. Calyx-tube (fig. 8/6, p. 28) cylindric-campanulate, 5–7 mm. long, 3–4 mm. in diameter, sparsely to densely pilose, the hairs intermixed with linear-subulate caducous appendages 1–2 mm. long and setose at the apex (variously branched hairs may be present along the length of the appendage); lobes subulate-lanceolate, 6 mm. long, margin ciliate, setose at the apex; intersepalar appendages ligulate, 2–2·5 mm. long, setose at the apex. Petals pink, obovate, 20 mm. long, 15 mm. wide. Stamens 10, unequal; large stamens with anthers 7 mm. long, connective produced 6 mm., bilobed appendage 1·5 mm. long, filament 6 mm. long; small stamens with anthers 5·5 mm. long, connective produced 1 mm., bilobed appendage 0·5 mm. long, filament 5 mm. long.

UGANDA. Bunyoro District: Hoima, Feb. 1943, *Purseglove* 1242!; Busoga District: Kibibi, 6 Feb. 1953, *G. H. Wood* 636!; Masaka District: Bugala I., July 1945, *Purseglove* 1727!
KENYA. Kwale District: Majoreni, Nov. 1929, *R. M. Graham* in *F.D.* 2189! & W. of Ramisi, 9 May 1953, *Bally* 8898! & Shimba Hills, Lango ya Mwagandi [Longomwagandi] Forest, 14 Nov. 1968, *Magogo & Estes* 1227!
TANZANIA. Tanga District: Magunga Estate, 14 Oct. 1952, *Faulkner* 1055!; Mpanda District: Kungwe Mt., Kasoje, 17 July 1959, *Newbould & Harley* 4431!; Uzaramo District: Vikindu Forest Reserve, Feb. 1952, *Procter* 33!; Zanzibar I., Mkokotoni, 17 Sept. 1960, *Faulkner* 2712!; Pemba I., Chake Chake, 26 July 1929, *Vaughan* 405!
DISTR. **U**2–4; **K**7; **T**1, 3, 4, 6–8; **Z**; **P**; widespread in tropical Africa from Sierra Leone southwards to Angola and extending eastwards through Zaire and E. Africa to Rhodesia and Mozambique; introduced into Malesia and the West Indies
HAB. Margins of rain-forest, riverine forest, flood plains and valley grassland, swamps, upland grassland in moist places; sea-level to 1900 m.

SYN. *Osbeckia rotundifolia* Sm. in Rees, Cycl. 25, *Osbeckia* (1822)
Melastoma plumosa D. Don in Mem. Wern. Nat. Hist. Soc. 4: 291 (1823). Type: Sierra Leone, *Afzelius* (BM, holo.!)

M. prostrata Schumach., Beskr. Guin. Pl.: 220 (1827) & in Kongel. Dansk. Vid. Selsk. Nat. Math. Afh. 3: 240 (1828). Type: Ghana, Aquapim, *Thonning* (C, holo.)
Heterotis plumosa (D. Don) Benth. in Hook., Niger Fl.: 348 (1849)
H. prostrata (Schumach.) Benth. in Hook., Niger Fl.: 349 (1849)
Osbeckia zanzibariensis Naud. in Ann. Sci. Nat., sér. 3, 13: t. 7/5 (1850) & 14: 55 (1850). Type: Zanzibar, *Bojer* (P, holo., K, iso.!)
Lepidanthemum triplinervum Klotzsch in Peters, Reise Mossamb., Bot. 1: 64 (1861). Type: Mozambique, Boror, Apr. 1846, *Peters* (B, holo. †)
Dissotis plumosa (D. Don) Hook. f. in F.T.A. 2: 452 (1871)
D. prostrata (Schumach.) Hook. f. in F.T.A. 2: 452 (1871); Triana in Trans. Linn. Soc. 28: 58 (1872)
D. rotundifolia (Sm.) Triana var. *prostrata* (Schumach.) Jacques-Félix in Adansonia 11: 548 (1971)

NOTE. The wide range of variation found in the species is believed to be due to habitat conditions and the differences consequently do not warrant taxonomic recognition.

7. **D. canescens** (*Graham*) *Hook. f.* in F.T.A. 2: 453 (1871); Jacques-Félix in Bull. I.F.A.N. 15: 979 (1953); F.W.T.A., ed. 2, 1: 258 (1954); A. & R. Fernandes in Mem. Soc. Brot. 11: 26, 80 (1956) & in Bol. Soc. Brot., sér. 2, 34: 186 (1960) & in C.F.A. 4: 145 (1970). Type: grown at Royal Botanical Garden, Edinburgh, from material of unknown origin supplied by Berlin (E, holo.)

Woody herb 50–120(–180) cm. high, usually erect, sometimes decumbent and rooting at the lower nodes; stem stellate-pubescent. Leaf-lamina oblong-lanceolate to narrowly oblong-ovate, 1·7–8·5 cm. long, 0·3–2·2 cm. wide, apex obtuse to acute, base rounded, stellate-pubescent above, hoary-stellate beneath; midrib and 1 or 2 pairs of basal longitudinal nerves impressed above, prominent beneath; petiole 0·5 mm. long. Inflorescence of long leafy many-flowered panicles; flowers 5-merous; pedicel 1–2 mm. long. Calyx-tube (fig. 8/7, p. 28) campanulate, 4·5 mm. long, 2·5–3 mm. in diameter, stellate-pubescent or stellate-pubescent mixed with simple or capitate hairs or minute ciliate appendages or rarely with simple or capitate hairs only; lobes triangular-lanceolate, 4–5 mm. long, persistent, ± densely pubescent; intersepalar appendages usually absent in the Flora area, rarely as a bristle 0·2–0·5 mm. long or as a caducous ligulate appendage 1–2 mm. long. (Careful examination under low power of a microscope is necessary to find these intersepalar appendages as well as to determine the form of the indumentum.) Petals reddish purple, broadly obovate, 15 mm. long, 11 mm. wide, margin glabrous or ciliate. Stamens 10, unequal; 5 large stamens with anthers 5·5 mm. long, connective produced 8 mm., clavate apically emarginate appendage 1·5 mm. long, filament 6 mm. long; 5 short stamens with anthers 5 mm. long, free part of connective 1 mm. long, bilobed appendages 0·5 mm. long, filament 5·5 mm. long. Calyx accrescent in fruit, tube enlarging to 7 mm. long, 5 mm. in diameter, somewhat glabrescent; capsule hoary at the apex.

UGANDA. Ankole District: Igara Mitoma, *Purseglove* 496!; Busoga District: Iganga, Walugogo A.L.G. Plantation, 13 Aug. 1952, *G. H. Wood* 323!; Masaka District: Katera, 1 Oct. 1953, *Drummond & Hemsley* 4511!
KENYA. Trans-Nzoia District: Kitale, 16 Sept. 1956, *Bogdan* 4293!; Uasin Gishu District: Kipkarren, Aug. 1931, *Brodhurst-Hill* 87!; N. Kavirondo District: Elgon, SW. slopes, Dec. 1935, *Tweedie* 134!
TANZANIA. Mwanza District: NW. Uzinza, Buchosa-Bugando Chiefdoms, 16 June 1937, *B. D. Burtt* 6585!; Buha District: 150 km. Kibondo–Kasulu, 15 July 1960, *Verdcourt* 2828!; Songea District: Mpapa, 18 Oct. 1956, *Semsei* 2605!
DISTR. **U**1–4; **K**3, 5; **T**1, 4, 5, 7, 8; also Nigeria, Cameroun, Central African Republic, Zaire, Ethiopia, and southwards through Mozambique and Angola to the Cape
HAB. Valley grassland, margins of swamp forest and marshy ground; 790–1950 m.

SYN. *Osbeckia canescens* Graham in Edin. New Phil. Journ. 28: 399 (1840) & in Bot. Mag. 66, t. 3790 (1840)

O. umlaasiana Hochst. in Flora 27: 424 (1844); Sond. in Fl. Cap. 2: 518 (1862). Type: South Africa, Natal, Umlaas, *Krauss* 120 (B, holo. †, BM, K, iso.!)

O. incana Walp., Repert. Bot. Syst. 5: 708 (1846), *nom. illegit.* Type: as for *O. umlaasiana* Hochst.

Argyrella incana (Walp.) Naud. in Ann. Sci. Nat., sér. 3, 13: 300, t. 6/7 (1850), *nom. illegit.*

A. canescens (Graham) Harv., Gen. S. Afr. Pl.: 113 (1868)

Dissotis incana (Walp.) Triana in Trans. Linn. Soc. 28: 58, t. 4/44d (1872); Cogn. in A. & C. DC., Monogr. Phan. 7: 370 (1891); Taub. in P.O.A. C: 295 (1895); Gilg in E.M. 2: 17 (1898); V.E. 3 (2): 749 (1921), *nom. illegit.*

Tristemma verdickii De Wild. in Ann. Mus. Congo, Bot., sér. 4, 1: 219 (1903), *non Dissotis verdickii* De Wild. (1903). Type: Zaire, Katanga, Lukafu, *Verdick* 621a (BR, holo.)

NOTE. Jacques-Félix, in Bull. I.F.A.N. 15: 980 (1953), recognizes a widespread var. *canescens* with a stellate-pubescent calyx-tube and a var. *sudanense* Jacques-Félix with simple hairs only and restricted to Nigeria, Cameroun and the Central African Republic. A. & R. Fernandes, in An. Junta Invest. Ultramar 10: 31 (1955) & in Mem. Soc. Brot. 11: 27 (1956), have attempted to differentiate between entities having stellate hairs only on the calyx-tube (var. *canescens* sensu stricto) and those with a mixture of stellate and capitate hairs (var. *zambeziensis, nom. inval.*), but due to the presence of intermediate forms concluded that varieties could not be maintained. A careful examination of the East African material shows the full range of variation between stellate hairs only, a mixture of stellate hairs and ciliate appendages to capitate hairs only; furthermore, the length of the calyx is not correlated with the indumentum, as previously suggested. Mixed populations have also been noted. Further field work and breeding studies are necessary before infraspecific variation can be understood in this species.

8. **D. brazzae** *Cogn.* in A. & C. DC., Monogr. Phan. 7: 372 (1891); Jacques-Félix in Bull. Mus. Hist. Nat., sér. 2, 7: 370 (1935); F.W.T.A., ed. 2, 1: 258 (1954); A. & R. Fernandes in Mem. Soc. Brot. 11: 34, 85, t. 6 (1956) & in C.F.A. 4: 148 (1970). Type: Gabon, Franceville, *Brazza* (P, holo.)

Erect perennial herb 0·6–1·5 m. high; stem sharply 4-angled or winged, appressed setose. Leaf-lamina narrowly ovate, 2–10 cm. long, 0·7–4·5 cm. wide, acuminate, base truncate to subcordate, margin serrate with ± appressed bristles, shortly and lightly appressed pilose above, more densely so beneath; midrib and 3 pairs of basal longitudinal nerves impressed above, prominent beneath; petiole 1·5–6(–10) mm. long. Inflorescence lax axillary cymes forming a leafy terminal panicle; flowers 5-merous. Calyx-tube (fig. 9/1, p. 30) cylindric-urceolate, 4·5 mm. long, 3 mm. in diameter, setose and with caducous stalked peltate scales that are compound stellate-setose; calyx-lobes triangular-ligulate, 3 mm. long, caducous; intersepalar appendages absent. Petals reddish purple, obovate, 17 mm. long, 12 mm. wide. Stamens 10, unequal; long stamens with anthers 8 mm. long, connective produced 3 mm., bilobed clavate appendage 2 mm. long, filament 6 mm. long; short stamens with anthers 7 mm. long, connective produced 1·5 mm. long, bilobed appendage 1 mm. long, filament 6 mm. long; "osbeckioid" forms with 10 short stamens also occur in the Flora area. Neck of calyx-tube elongating in fruit, the tube becoming 8 mm. long, sparsely setose; capsule setose at the apex.

UGANDA. Ankole District: Ruizi R., 16 Mar. 1951, *Jarrett* 412!; Busoga District: Kyemeire, 28 Oct. 1952, *G. H. Wood* 492!; Mengo District: Mbuya hill, 3 Aug. 1969, *Rwaburindore* 76!

KENYA. Nandi District: Kaimosi, 15 June 1953, *G. R. Williams* 570!; N. Kavirondo District: Kakamega Forest, 15 Oct. 1953, *Drummond & Hemsley* 4780! & 22 Dec. 1967, *Perdue & Kibuwa* 9425!

TANZANIA. Mwanza District: Ukerewe I., Rubya Forest Reserve, 22 June 1958, *Makwilo Semkiwa* 50!; Buha District: Manyovu, July 1955, *Procter* 427!; Ulanga District: Sali, 17 Mar. 1932, *Schlieben* 1899!

DISTR. **U**2–4; **K**3, 5; **T**1, 3, 4, 6–8; Sierra Leone, Guinée, Togo, Cameroun, Zaire, Rwanda, Gabon, Zambia and Angola
HAB. Forest margins, valley and upland grassland and deciduous bushland; 900–1800 m.

SYN. [*D. multiflora* sensu Gilg in E.M. 2: 18, t. 2/F (1898) & V.E. 3 (2): 751 (1921), pro parte quoad syn. *D. brazzae* Cogn.; T.T.C.L.: 309 (1949), *non* (Sm.) Triana]
D. tanganyikae Kraenzlin in Viert. Nat. Ges. Zürich 76: 150 (1931). Type: Zaire, Katanga, Kibandu, *Kassner* 3055a (Z, holo.)

9. **D. caloneura** *Engl.*, V.E. 3 (2): 749 (1921); A. & R. Fernandes in Bol Soc. Brot., sér. 2, 30: 171, t. 3, 4 (1956) & in Mem. Soc. Brot. 11: 77 (1956) & in Bol. Soc. Brot., sér. 2, 34: 59 (1960). Type: Zaire, Katanga, Nunta Mt., *Kassner* 2960 (B, holo. †, BM, fragment!)*

Small tree 1·5–5 m. high; bark pale grey, corky, peeling in strips; branchlets 4-angular, sparsely to densely setose, becoming glabrescent. Leaf-lamina lanceolate, ovate-lanceolate to ovate, 5–10 cm. long, 3–7·5 cm. wide subacuminate, base subcordate, margin dentate, setose-ciliate, bullate with sparse short bulbous-based setae, or with dense long bulbous-based setae above, sparsely or densely setose beneath; midrib and 2 pairs of basal longitudinal nerves impressed above, prominent beneath; reticulation subprominent or prominent; petiole 1–2 cm. long. Inflorescence a many-flowered lax or congested terminal paniculate-cyme; bracts caducous; pedicels 2–3 mm. long. Calyx-tube (fig. 9/2, p. 30) urceolate, 9–10 mm. long, 5 mm. in diameter, glabrous (or strigose, but not in the Flora area); lobes persistent, ovate, 4 mm. long, 4 mm. wide, margin ciliate, otherwise glabrous. Petals mauve, trapeziform, ± 18 mm. long, 16 mm. wide. Stamens 10, markedly unequal; 5 large stamens with anthers 10 mm. long, free part of connective arcuate, 8 mm. long, appendage clavate, 1·5 mm. long, filament 11 mm. long; 5 small stamens with anthers 8 mm. long, connective produced 0·5 mm., appendage 1 mm. long, filament 7 mm. long. Fruiting calyx enlarging to 12 mm. long, 6 mm. in diameter; calyx-lobes reflexed; capsule not exserted.

Four varieties are recognized, 3 of which occur in the Flora area.

KEY TO INFRASPECIFIC VARIANTS

Leaf-lamina with sparse short (± 0·5 mm.) setae:
  Inflorescence lax . . . . var. **caloneura**
  Inflorescence congested . . . . . . var. **confertiflora**
Leaf-lamina with dense long (± 1·5 mm.) setae . var. **setosior**

var. **caloneura**; A. & R. Fernandes in Bol. Soc. Brot., sér. 2, 34: 59 (1960)

Leaf-lamina up to 7·5 cm. long, 4 cm. wide, sparsely setose, setae ± 0·5 mm. long. Inflorescence a lax paniculate cyme.

TANZANIA. Ufipa District: Kalambo Falls, 26 Mar. 1960, *Richards* 12796!
DISTR. **T**4; also in Zaire, Burundi and Zambia
HAB. Among rocks; 1200 m.

SYN. *D. venulosa* Hutch., Botanist in S. Afr.: 512 (1946). Type: Zambia, Mbala District, Lake Chila, *Hutchinson & Gillett* 3887 (K, holo.!)

var. **confertiflora** *A. & R. Fernandes* in Bol. Soc. Brot., sér. 2, 34: 59, t. 1 (1960) & in 34: 185 (1960). Type: Tanzania, Ufipa District, Kito [Keto] Mt., *Richards* 6176 (K, holo.!, COI, iso.!)

Similar to var. *caloneura* but with a densely congested almost globose inflorescence.

* The specimens of *Kassner* 2960 at BM, K and P are *D. trothae*; the BM specimen also includes a single flower of *D. caloneura* contained in a packet.

TANZANIA. Ufipa District: Kito Mt., 13 Sept. 1956, *Richards* 6176
DISTR. **T4**; not known elsewhere
HAB. Among rocks at summit of mountain

NOTE. Further material and field studies are required to determine whether the form of the inflorescence is due simply to the wind-swept nature of the type habitat.

var. **setosior** *A. & R. Fernandes* in Bol. Soc. Brot., sér. 2, 30: 171, t. 5 (1956) & in Mem. Soc. Brot. 11: 77 (1956). Type: Tanzania, Kigoma District, Kafulu, *Eggeling* 6172 (EA, holo., K, iso.!)

Leaf-lamina up to 10 cm. long, 7·5 cm. wide, densely setose above, setae ± 1·5 mm. long; reticulation prominent beneath and densely hoary setose. Inflorescence a lax paniculate cyme.

TANZANIA. Kigoma District: Kafulu, July 1951, *Eggeling* 6172!
DISTR. **T4**; not known elsewhere
HAB. Sandstone outcrops, with *Parinari* and *Faurea*; 1500 m.

10. **D. arborescens** *A. & R. Fernandes* in Bol. Soc. Brot., sér. 2, 29: 51, t. 6 (1955) & in Mem. Soc. Brot. 11: 25, 78 (1956) & in Bol. Soc. Brot., sér. 2, 34: 185 (1960). Type: Tanzania, Iringa District, Sao Hill, *Greenway* 6176 (EA, holo., K!, PRE, iso.)

Small tree or shrub 1–6 m. high, flowering before the leaves appear; bark grey-brown, fissured and eventually peeling to reveal a yellowish brown cortex; branchlets woody, deeply sulcate between the nodes when young, densely yellow-brown appressed setose, becoming glabrescent. Leaf-lamina elliptic-ovate, 4·5–12 cm. long, 2·3–8·5 cm. wide, apex acute to shortly acuminate, base cuneate to subcordate, margin minutely denticulate, bullate and densely appressed setose above, more finely setose on the venation beneath; midrib and 2 pairs of basal longitudinal nerves impressed above, prominent beneath, reticulation prominent; petiole 1–3·6 cm. long. Flowers 5-merous, ± 6 densely clustered at the apices of the branches; pedicels stout, ± 1 mm. long; bracts caducous, lanceolate to broadly ovate, ± 10 mm. long, 7·5 mm. wide, densely appressed setose outside. Calyx-tube (fig. 9/6, p. 30) cup-shaped, 8 mm. long, 8 mm. in diameter, densely appressed pilose when young, becoming more sparsely pilose, with a ± glabrous apical rim; lobes oblong-ovate, 7–8 mm. long, 6 mm. wide, apex rounded or obliquely truncate, margins ciliate, otherwise glabrous; intersepalar appendages absent. Petals magenta, obcordate, 22–26 mm. long, 19–23 mm. wide, apex emarginate, margin shortly ciliate. Stamens 10, markedly unequal; 5 large stamens with anthers 11 mm. long, connective produced 9–14·5 mm., appendage 1·5–2·5 mm. long, incurved, filament 12 mm. long; 5 small stamens with anthers 9 mm. long, connective produced 1·5 mm., appendage ± 2·5 mm. long, somewhat incurved, filament 9 mm. long. Fruiting calyx enlarging to 10 mm. long, 8 mm. in diameter, calyx-lobes reflexed; capsule slightly exserted, apex setose.

TANZANIA. Iringa District: near Sao Hill, 30 Oct. 1947, *Brenan & Greenway* 8243! & 85 km. S. of Iringa, 14 July 1963, *Mathias & Taylor* A92! & James Corner, 10 Aug. 1966, *Gillett* 17395!
DISTR. **T7**; not known elsewhere
HAB. Fringes of upland rain-forest, upland grassland; 1500–2340 m.

SYN. *D. sp.* sensu T.T.C.L.: 310 (1949)

11. **D. glandulicalyx** *Wickens* in K.B. 29: 142, fig. 1 (1974). Type: Tanzania, Mpanda District, Kungwe Mt., *Harley* 9542 (K, holo.!)

Small tree or shrub 1·3–3·6 m. high; bark pale grey, fissured and eventually peeling to reveal a purplish red cortex; branchlets woody, deeply sulcate between the nodes when young, densely reddish brown setose, becoming glabrescent. Leaf-lamina elliptic, 4–15 cm. long, 2·7–10 cm. wide, apex acute to shortly acuminate, base subcordate, pilose and bullate above, pilose on the reticulation beneath; midrib and 2 pairs of basal longitudinal nerves scarcely visible above, prominent beneath, reticulation prominent beneath; petiole 1·5–2·5 cm. long. Inflorescence a terminal panicle; flowers 5-merous, produced when the tree is leafless; pedicels stout, 3–4 mm. long; bracts caducous, lanceolate, 5 mm. long, 2 mm. wide, densely setose. Calyx-tube (fig. 9/7, p. 30) cylindric-campanulate, 10–11 mm. long, 6 mm. in diameter, densely covered with glandular-setose tufts; lobes oblong-ovate, 4 mm. long, 2·7 mm. wide, apex rounded, margin ciliate, central portion of lobe with glandular-setose tufts and glandular-setose scales, muricate towards the margins; intersepalar appendages ligulate, 1·5 mm. long, glandular-setose. Petals pink to purple, obovate, 22 mm. long, 17 mm. wide, margin shortly ciliate, otherwise glabrous. Stamens 10, markedly unequal; 5 large stamens with anthers 11 mm. long, connective produced 13 mm., bilobed appendage 2 mm. long, filament 12 mm. long; 5 small stamens with anthers 9 mm. long, connective produced 1·5 mm., bilobed appendage 1·2 mm. long, appressed to the connective, filament 8·5 mm. long. Fruiting calyx enlarging to 12 mm. long, 7 mm. in diameter; calyx-lobes reflexed; capsule slightly exserted, apex setose.

TANZANIA. Mpanda District: Mahali Mts., Sisaga, 24 Aug. 1958, *Newbould & Jefford* 1772! & 30 Aug. 1958, *Newbould & Jefford* 1947! & Musenabantu, 14 Aug. 1959, *Harley* 9352!

DISTR. **T4**; not known elsewhere

HAB. Upland grassland; 1770–2340 m.

12. **D. melleri** *Hook. f.* in F.T.A. 2: 453 (1871)*; Triana in Trans. Linn. Soc. 28: 58 (1872); Cogn. in A. & C. DC., Monogr. Phan. 7: 371 (1891); Taub. in P.O.A. C: 295 (1895); Gilg in E.M. 2: 18 (1898); V.E. 3 (2): 750 (1921); A. & R. Fernandes in Bol. Soc. Brot., sér. 2, 30: 172, t. 6, 7 (1956), descr. ampl., & in Mem. Soc. Brot. 11: 25, 79 (1956) & in Bol. Soc. Brot., sér. 2, 34: 185 (1960). Type: Malawi, Mt. Chiradzura, Manganja range, Sept. 1861, *Meller* (K, holo.!)

Small tree or shrub 2–4·5 m. high; branchlets 4-angular, appressed setose. Leaf-lamina elliptic-lanceolate to elliptic, 4–12 cm. long, 1·7–5·5 cm. wide, apex subacute to subacuminate, base truncate to subcordate, densely to thinly appressed setose above (setae bulbous-based, ± 2 mm. long), shortly setose beneath; midrib and 2 or 3 pairs of basal longitudinal nerves impressed above, prominent beneath; petiole 6–20 mm. long. Inflorescence a many-flowered racemose cyme sometimes forming a terminal panicle; bracts caducous; pedicels ± 2 mm. long, thinly setose or glandular pubescent. Calyx-tube (fig. 9/4, 5, p. 30) urceolate, 8–9 mm. long, 6 mm. in diameter, covered with small simple setae mixed with setose appendages and sometimes with glandular setae (var. *greenwayi*); lobes persistent, broadly oblong, 5 mm. long, 6 mm. wide, margin ciliate, otherwise glabrous. Petals purple, obovate, 35 mm. long, 24 mm. wide. Stamens 10, markedly unequal; 5 large stamens with anthers 10–11 mm. long, free part of connective arcuate, 13 mm. long, appendage clavate, 2 mm. long, with posterior lobe, filament 14–15 mm. long; 5 small stamens with anthers 10 mm. long, connective produced 1 mm., appendage auriculate, 1·5 mm. long, filament 12 mm. long. Fruiting receptacle increasing to 10 mm. long, 7 mm. in diameter; capsule not exserted.

* Original description often attributed to Hook. f. ex Triana, but see Miss Raphael in Biol. J.L.S. 2: 73 (1970) for date of Triana's paper.

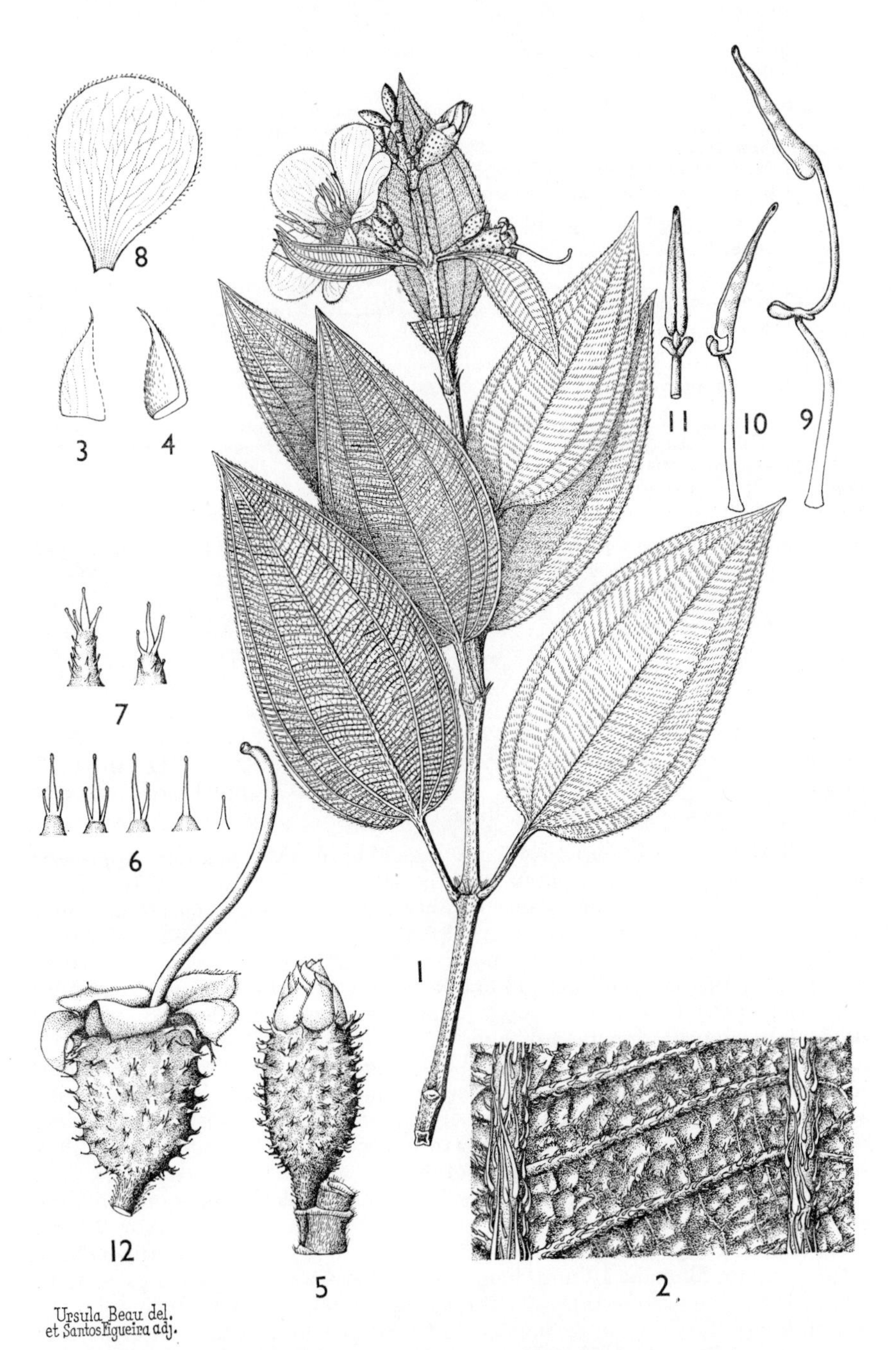

Fig. 13. *DISSOTIS MELLERI* var. *GREENWAYI*—**1,** flowering branch, × ½; **2,** part of leaf, lower surface, × 6; **3, 4,** bracteoles × 2; **5,** flower bud, × 2; **6,** setae and appendages from lower and middle parts of calyx-tube, × 6; **7,** appendages from upper part of calyx-tube, × 6; **8,** petal, × 1; **9,** large stamen, × 2; **10, 11,** small stamen, side and front views respectively, × 2; **12,** flower after fall of petals and stamens, × 2. All from *Greenway* 8410. Reproduced by permission of the Editors of Boletim da Sociedade Broteriana.

var. **melleri**

Petiole 6–7 mm. long; transverse reticulation on leaves obscure. Pedicel and calyx-tube eglandular. Fig. 9/5, p. 30.

TANZANIA. Ufipa District: Mbisi, 3 Oct. 1936, *Lea* L.U. 5!; Mbeya Peak Forest Reserve, 30 May 1962, *Kerfoot* 4249!; Songea District: Matengo Hills, about 21 Jan. 1901, *Busse* 921!
DISTR. **T4**, 7, 8; also in Malawi, Mozambique and Zambia
HAB. Upland rain-forest and upland evergreen bushland; 1800–3100 m.

SYN. *D. whytei* Bak. in K.B. 1897: 267 (1897). Type: Malawi, Mt. Zomba, Dec. 1896, *Whyte* (K, holo.!)

var. **greenwayi** (*A. & R. Fernandes*) *A. & R. Fernandes* in Bol. Soc. Brot., sér. 2, 46: 68 (1972). Type: Tanzania, Rungwe District, R. Kiwira, *Greenway* 8410 (EA, holo.!, K!, PRE, iso.)

Petiole 7–20 mm. long; transverse reticulation on leaves prominent. Pedicel and calyx with glandular setae. Fig. 9/4, p. 30, & 13, p. 45.

TANZANIA. Rungwe District: without precise locality, 4 Sept. 1932, *R. M. Davies* 239! & Poroto Mts., Matesse R. to Tukuyu, 27 Oct. 1947, *Brenan & Greenway* 8225! & Ngozi–Poroto Mts., 16 Oct. 1956, *Richards* 6464!
DISTR. **T7**; Malawi and Zambia (Nyika Plateau)
HAB. Riverine forest and margins of upland rain-forest; 1550–2100 m.

SYN. *D. rubro-violacea* Gilg in E.J. 30: 365 (1901) & in V.E. 3 (2): 749 (1921), *non* sensu Engl., V.E. 3 (2): 751 (1921); T.T.C.L.: 310 (1949). Type: Tanzania, N. Mt. Rungwe, *Goetze* 1136 (B, holo. †, BM, iso.!)
D. greenwayi A. & R. Fernandes in Bol. Soc. Brot., sér. 2, 30: 172, t. 8 (1956) & in Mem. Soc. Brot. 11: 25, 75 (1956) & in Bol. Soc. Brot., sér. 2, 34: 185 (1960)

13. **D. aprica** *Engl.*, V.E. 3 (2): 749 (1921); T.T.C.L.: 309 (1949); A. & R. Fernandes in Bol. Soc. Brot., sér. 2, 30: 174, t. 9 (1956), descr. ampl., & in Mem. Soc. Brot. 11: 80 (1956). Type: Tanzania, Lindi District, Rondo [Muera] Plateau, *Busse* 2577 (B, holo. †, EA, iso.!)*

Small tree or shrub up to 4·5 m. high; branchlets angular, appressed setose when young, becoming glabrescent. Leaf-lamina ovate-oblong, 4·5–13·5 cm. long, 2–6 cm. wide, membranous, acuminate, base cuneate or rounded, margin slightly serrate and long-ciliate, thinly setose above (setae ± 3 mm. long), sparsely setose beneath (setae up to 1–2 mm. long); midrib and 1 or 2 pairs of basal longitudinal nerves impressed above, subprominent beneath; petiole 1–3 cm. long. Inflorescences few-flowered terminal paniculate cymes; bracts caducous, 15–25 mm. long; pedicels ± 3 mm. long. Calyx-tube (fig. 9/3, p. 30) cylindric-campanulate, 10 mm. long, 6 mm. in diameter, densely clothed in setose appendages usually intermixed with simple setae; lobes persistent, broadly oblong, 6 mm. long, 5 mm. wide, margin ciliate, apex with a stellate-setose appendage; stellate-setose intersepalar appendages also present. Petals red or violet, obcordate, 28 mm. long, 25 mm. wide, margin ciliate. Stamens 10, markedly unequal; 5 large stamens, anthers 10–12 mm. long, free part of connective arcuate, 7 mm. long, anterior appendage oblong, ± 2 mm. long, posterior appendages bituberculate, filament 10 mm. long; 5 small stamens with anthers 10 mm. long, connective produced up to 0·5 mm., appendage ± 2 mm. long, filament 8 mm. long. Fruiting calyx increasing to 10 mm. long, 6 mm. in diameter; capsule not exserted.

* The collector is not cited in the original description, but as Busse is believed to have been the only collector on the Rondo Plateau before 1920 there can be little doubt that *Busse* 2577 is an isotype.

TANZANIA. Kilosa District: Ukaguru Mts., Mamiwa Forest Reserve, near Mandege, 8 Aug. 1972, *Mabberley* 1421 !; Lindi District: Rondo Plateau, 16 May 1903, *Busse* 2577 ! & Feb. 1951, *Eggeling* 6051 & Rondo Plateau, Mchinjiri, Mar. 1952, *Semsei* 696 !

DISTR. T6, 8; not known elsewhere

HAB. Margins of rain-forest and semi-deciduous forest; 900–1600 m.

NOTE. The Mabberley specimen from the Ukaguru Mts. differs slightly from those collected on the Rondo Plateau in having a much finer indumentum throughout and without simple setae intermixed with the setose appendages on the calyx.

14. **D. bussei** *Engl.*, V.E. 3 (2): 749 (1921); T.T.C.L.: 309 (1949); A. & R. Fernandes in Bol. Soc. Brot., sér. 2, 29: 53, t. 7 (1955), descr. ampl., & in Mem. Soc. Brot. 11: 78 (1956) & in Bol. Soc. Brot., sér. 2, 34: 186 (1960). Type: Tanzania, Mpwapwa District, Kiboriani Mts., collector not cited, presumably *Busse** (B, holo. †)

Small tree or shrub 2–5 m. high, flowering shortly before the leaves appear; bark grey, fissured; branches woody, sulcate, nodes swollen, young shoots 4-angled, yellowish and densely setose. Leaf-lamina broadly ovate, 7–21 cm. long, 5–18 cm. wide, apex obtuse, base cordate, margin entire, green or reddish and setose-bullate above, densely greyish yellow tomentose beneath; midrib and 2–3 pairs of basal longitudinal nerves impressed above, prominent beneath, reticulation prominent beneath; petiole 1–3·5 cm. long. Inflorescence cymose, ± 15–25- or more flowered; pedicels ± 3–8 mm. long, setose; bracts caducous, ± 4 mm. long, spinulose-setose. Calyx-tube (fig. 9/8, p. 30) cylindric-campanulate, 7–9 mm. long, 5–7 mm. in diameter, densely beset by minute bulbous-based setae 0·5 mm. long; lobes persistent, obliquely-deltoid, 4–4·5 mm. long, 3·5–4 mm. wide, reddish, sparsely setose, margin ciliate. Petals purple, broadly obovate, ± 22 mm. long, 19 mm. wide, margin ciliate. Stamens 10, markedly unequal; 5 large stamens with anthers 11 mm. long, free part of connective arcuate, 10 mm. long, clavate appendage ± 2 mm. long, filament 12 mm. long; 5 small stamens with anthers 10 mm. long, connective produced 1 mm., appendage ± 1·6 mm. long and apically dilated, filament 10 mm. long. Fruiting calyx enlarging to 10 mm. long and 7 mm. in diameter, calyx-lobes reflexed; ovary not exserted, apex setose.

TANZANIA. Kondoa District: Salanga Mt., 15 Oct. 1932, *B. D. Burtt* 4412 !; Mpwapwa, 20 Aug. 1930, *Greenway* 2442 !; Iringa District: Iliangelo [Kiangole] Ridge, July 1953, *Carmichael* 230 !

DISTR. T5, 7; not known elsewhere

HAB. Upland rain-forest margins and upland deciduous woodland; 1350–2100 m.

SYN. [*D. rubro-violacea* sensu A. & R. Fernandes in Mem. Soc. Brot. 11: 78 (1956), *non* Gilg (1902)]
D. sp. no. 16 sensu T.T.C.L.: 310 (1949)

15. **D. dichaetantheroides** *Wickens* in K.B. 29: 141 (1974). Type: Tanzania, Morogoro District, Nguru Mts., *Drummond & Hemsley* 1960 (K, holo. !, EA, SRGH, iso.)

Shrub 2–3 m. high; branchlets ± densely and shortly appressed setose. Leaf-lamina oblong-ovate, 3·5–9 cm. long, 1·5–4 cm. wide, apex acute, base rounded to shortly cuneate, sparse short curved setae above, sparse longer but straight setae beneath; midrib and 2 pairs of longitudinal nerves impressed above, subprominent beneath; petiole 1–2 cm. long, ± densely appressed setose. Inflorescence few–many-flowered lax terminal panicle extending beyond the leaves; flowers 5-merous; pedicels 3 mm.; bracts minute, ± 1 mm. long. Calyx-tube campanulate, 4–4·5 mm. long, 4 mm. in

* Busse collected in the Kiboriani Mts. in September 1900, Nos. 280–300.

diameter, sparsely setose (setae ± 0·5 mm. long); lobes rounded to trapeziform, 1·5 mm. long, 2 mm. wide, glabrous except for the shortly ciliate margin, persistent; intersepalar ligulate appendages tardily deciduous, ± 0·5 mm. long, with 2–3 setae at the apex. Petals obovate, 14 mm. long, 10 mm. wide, probably larger in mature flowers. Stamens 10, markedly unequal; 5 large stamens with anthers 6 mm. long, connective produced 3 mm., slightly arcuate, anterior appendage with 2 clavate lobes 1·5 mm. long, extended posteriorly into a single spur 1 mm. long, filament 8–9 mm. long; 5 short stamens with anthers 5 mm. long, connective produced 1 mm., clavately bilobed appendage 1 mm. long, auricled at the base, filament 7 mm. long. Fruiting capsule enlarging to 6 mm. long, 4·5 mm. in diameter; capsule setose at the apex, not exserted.

TANZANIA. Morogoro District: Nguru Mts., 2 Apr. 1953, *Drummond & Hemsley* 1960! & Nguru South Forest Reserve, 5 Feb. 1971, *Mabberley* 687!
DISTR. **T6**; not known elsewhere
HAB. Upland rain-forest; 1450–2000 m.

SYN. [*D. polyantha* sensu A. & R. Fernandes in Bol. Soc. Brot., sér. 2, 34: 185 (1960), *non* Gilg (1921)]
[*Dichaetanthera corymbosa* sensu A. & R. Fernandes in Bol. Soc. Brot., sér. 2, 43: 297 (1969), *non* (Cogn.) Jacques-Félix]

16. **D. polyantha** *Gilg* in E.M. 2: 16, t. 2/E (1898) & in V.E. 3 (2): 749 (1921); T.T.C.L. 310 (1949); A. & R. Fernandes in Mem. Soc. Brot. 11: 77, t. 2 (1956). Type: Tanzania, Morogoro District, Kifuru, *Stuhlmann* 9081 (B, holo. †, K, photo.!)

Shrub or small tree 1·5–3 m. high; branchlets angular, densely hispid when young, the hairs reddish brown, bulbous-based. Leaf-lamina ovate or elliptic, 4–12·5 cm. long, 2·5–5·5 cm. wide, apex acute and shortly apiculate, base cuneate-rounded to subcordate, bullate and densely strigillose with bulbous-based hairs above, densely strigillose beneath; midrib and 1 pair of basal longitudinal nerves impressed above, prominent beneath, reticulation prominent beneath; petiole 5–40 mm. long, hispid. Inflorescence 5–12-flowered paniculate cymes; bracts caducous; pedicels 5–8 mm. long. Calyx-tube (fig. 9/9, p. 30) cylindric-campanulate, 5–7 mm. long, 4·5–5 mm. in diameter, densely covered with simple hairs thickened at the base; lobes persistent, ovate, 4–5 mm. long, 2–3 mm. wide, margin ciliate, otherwise glabrous. Petals reddish violet, obovate, ± 20 mm. long, 11 mm. wide. Stamens 10, markedly unequal; 5 large stamens with anthers 6·5 mm. long, free part of connective arcuate, 10 mm. long, appendage with 2 clavate lobes 2·5 mm. long, filament 7 mm. long; 5 small stamens with anthers 4 mm. long, connective produced 1 mm., clavately bilobed appendage 1·5 mm. long, filament 7 mm. long. Fruiting calyx enlarging to 8 mm. long, 7 mm. in diameter, calyx-lobes reflexed; setose apex of capsule not exserted.

TANZANIA. Morogoro District: Uluguru Mts., Lupanga Peak, *B. D. Burtt* 3467 & Uluguru Mts., above Morningside, Jan. 1953, *Eggeling* 6475!; Iringa District: Gologolo Mts., 13 Sept. 1970, *Thulin & Mhoro* 966!
DISTR. **T6, 7**; not known elsewhere
HAB. Upland dry evergreen forest and upland evergreen bushland; 1500–2100 m.

17. **D. senegambiensis** (*Guill. & Perr.*) *Triana* in Trans. Linn. Soc. 28: 58 (1872); A. & R. Fernandes in Garcia de Orta 2: 176, t. 8, & 284 (1954) & in Bol. Soc. Brot., sér. 2, 46: 69 (1972). Type: Senegal, Mboro, May 1826, *Leprieur* (P, holo.)

Woody perennial herb 0·3–2 m. high; branchlets hispid-pilose to scabrid. Leaves subsessile; lamina linear-lanceolate, oblong-lanceolate or oblong-ovate, 1·5–9·5(–11·5) cm. long, 0·5–3 cm. wide, apex acute, base subcuneate, margin slightly serrate and appressed setose, long setose-pilose on both surfaces; midrib and 1–2 pairs of basal longitudinal nerves impressed above, slightly raised beneath. Inflorescence of dense paniculate clusters; flowers 4–5-merous. Calyx-tube (fig. 10/1, 2, p. 31) ovoid-urceolate, 6 mm. long, 4 mm. in diameter, covered with stellate-setose linear or obtriangular capitate appendages, setae sometimes very dense and obscuring the flower; lobes caducous, narrowly triangular, 5 mm. long, margin shortly ciliate, apex with a single stellate-setose scale, otherwise glabrous; intersepalar appendages resembling those of the calyx-tube. Petals pinkish to purple, oblong-obovoid, 10–15 mm. long, 8–13 mm. wide, apex truncate or rounded. Stamens 8 or 10, markedly unequal or subequal in " osbeckioid " forms; large stamens with anthers 7–8 mm. long, free part of connective arcuate, 8·5 mm. long, bilobed appendage 2 mm. long, filament 8 mm. long; short stamens (including all of " osbeckioid " forms) with anthers 5–8 mm. long, connective produced 0·5–0·8 mm., bituberculate appendage 0·5 mm. long, filament 6 mm. long. Ovary with an apical tuft of setae, concealed by the calyx-tube. Fruiting calyx enlarging, mainly by an elongation of the neck, to 7–10 mm. long, 4·5 mm. in diameter; capsule not exserted.

var. **senegambiensis**; A. & R. Fernandes in Bol. Soc. Brot., sér. 2, 46: 69 (1972)

Flowers 5-merous. Stamens subequal (forma *senegambiensis*), or markedly unequal (forma *irvingiana* (Hook.) A. & R. Fernandes).*

UGANDA. Karamoja District: Mt. Kadam, Apr. 1959, *J. Wilson* 727 !; Ankole District: Kalinzu Forest, 21 May 1961, *Symes* 732 !; Mengo District: Entebbe, Sept. 1922, *Maitland* 215 !

KENYA. Nakuru, 7 Dec. 1956, *Verdcourt* 1618 !; Machakos District: Nzaui Hill, 18 June 1966, *D. Wood* 712 !; Kericho District: Amala [Mara] R., Aug. 1960, *Kerfoot* 2153 !

TANZANIA. Bukoba District: Mshamba, Dec. 1931, *Haarer* 2343 !; Kondoa, 16 June 1958, *Mahinde* 1 !; Iringa District: Dabaga Highlands, Idewa, 22 Feb. 1962, *Polhill & Paulo* 1568A !

DISTR. **U**1–4; **K**2–6; **T**1, 3, 5, 7; widespread in tropical Africa from Senegal to Ethiopia and southwards to Mozambique

HAB. Valley grassland, clearings and margins of upland rain-forest, upland grassland and moor, especially in damp places; 1050–2700 m.

SYN. *Osbeckia senegambiensis* Guill. & Perr. in Fl. Seneg. Tent. 1: 310 (1833); Hook. f. in F.T.A. 2: 443 (1871); Gilg in E.M. 2: 8 (1898); V.E. 3 (2): 746 (1921); F.P.S. 1:194 (1950); F.W.T.A., ed. 2, 1: 249 (1954)

Dissotis irvingiana Hook. in Bot. Mag. 85, t. 5149 (1859); Hook. f. in F.T.A 2: 453 (1871); Gilg in E.M. 2: 20 (1898); V.E. 3 (2): 752 (1921); F.W.T.A., ed. 2, 1: 259 (1954); A. & R. Fernandes in Garcia de Orta 2: 179 (1954). Types: Nigeria, Abeokuta, *Irving* 119 & Nupe, *Barter* 1025 (both K, syn. !)

Osbeckia abyssinica Gilg in E.M. 2: 8 (1898); V.E. 3 (2): 746 (1921); F.P.S. 1: 194 (1950). Type: Ethiopia, Begemder, Carruta, 1863, *Schimper* 1437 (B, holo. †, BM, K, iso. !)

O. saxicola Gilg in E.M. 2: 7, t. 1/D (1898); V.E. 3 (2): 746, fig. 317/c (1921); F.P.S. 1: 194 (1950). Type: Sudan, Jur [Ghattas Zeriba], 1869, *Schweinfurth* 1494 (K, lecto. !)

[*O. densiflora* sensu Gilg in E.M. 2: 8 (1898), quoad specim. *Buchanan* 114, *non* Gilg sensu stricto]

O. crepiniana Cogn. in Ann. Mus. Congo, Bot., sér. 1, 1: 23, t. 12/10–11 (1898); V.E. 3 (2): 746 (1921). Type: Zaire, Zambi, *Dewèvre* (BR, holo.)

O. cogniauxiana De Wild., Pl. Bequaert. 1: 374 (1922). Type: Zaire, Angi, *Bequaert* 5793 (BR, holo.)

Dissotis kassnerana Kraenzlin in Viert. Nat. Ges. Zürich 76: 151 (1931). Type: Kenya, Kitui District, Galunka, *Kassner* 861 (BM, K, iso. !)

D. irvingiana Hook. var. *irvingiana* forma *abyssinica* (Gilg) A. & R. Fernandes in Garcia de Orta 2: 179 (1954) & in Mem. Soc. Brot. 11: 36, 87 (1956) & in Bol. Soc. Brot., sér. 2, 34: 188 (1960).

* Forma *irvingiana* is only known from two doubtful collections in the Flora area; further gatherings of this forma are required in order to confirm its presence.

D. irvingiana Hook. var. *irvingiana* forma *irvingiana*; A. & R. Fernandes in An. Junta Invest. Ultramar 10, 3: 36 (1955) & in Mem. Soc. Brot. 11: 36 (1956) & in Bol. Soc. Brot., sér. 2, 34: 188 (1960)

var. **alpestris** (*Taub.*) *A. & R. Fernandes* in Bol. Soc. Brot., sér. 2, 46: 69 (1972). Type: Tanzania, Moshi District, Marangu, *Volkens* 631 (BM, lecto. !)

Flowers 4-merous; stamens markedly unequal (forma *alpestris*) or subequal (forma *osbeckioides* A. & R. Fernandes).*

UGANDA. Mbale District: W. Bugwe Forest Reserve, 20 May 1951, *G. H. Wood* 205! & Bugisu, Busano, 21 Jan. 1969, *Lye* 1690!

KENYA. Naivasha District: Hell's Gate, 15 Sept. 1965, *E. Polhill* 133!; S. Nyeri District: Kahuroini, 9 Dec. 1963, *Kibui* 33!; Masai District: Mt. Suswa, SE. flank, 30 Oct. 1962, *Glover & Samuel* [*Paulo*] 3383!

TANZANIA. Lushoto District: Sangerawe, 18 Sept. 1928, *Greenway* 859!; Tanga District: Ngua Forest Reserve, 22 Oct. 1962, *Semsei* 3552!; Uzaramo District: Pugu Forest Reserve, June 1954, *Semsei* 1770!

DISTR. **U**3; **K**3–6; **T**1–3, 6, 7; Ethiopia, Malawi and Mozambique

HAB. Valley grassland, forest edges, upland grassland and moor, especially on moist ground and around fumaroles; 900–2700 m.

SYN. *D. alpestris* Taub. in P.O.A. C: 295 (1895); Gilg in E.M. 2: 20 (1898); V.E. 3 (2): 752 (1921)

D. cincinnata Gilg in E.M. 2: 20 (1898); V.E. 3 (2): 752 (1921); T.T.C.L.: 309 (1949). Type: Tanzania, Moshi District, Marangu, *Volkens* 722 (B, holo. †)

D. irvingiana Hook. var. *alpestris* (Taub.) A. & R. Fernandes in Garcia de Orta 2: 179 (1954) & in Mem. Soc. Brot. 11: 37, 89 (1956) & in Bol. Soc. Brot., sér. 2, 34: 189 (1960)

D. irvingiana Hook. var. *alpestris* forma *alpestris*; A. & R. Fernandes, in An. Junta Invest. Ultramar 10, 3: 36 (1955) & in Mem. Soc. Brot. 11: 89 (1956) & in Bol. Soc. Brot., sér. 2, 34: 189 (1960)

D. irvingiana Hook. var. *alpestris* forma *osbeckioides* A. & R. Fernandes in An. Junta Invest. Ultramar 10, 3: 38 (1955); Type: Mozambique, Manica–Vumba, *Pedro & Pedróqão* 6989 in part (COI, holo., LMJ, iso.)

18. **D. densiflora** (*Gilg*) *A. & R. Fernandes* in Garcia de Orta 2: 180 (1954) & in Bol. Soc. Brot., sér. 2, 34: 189 (1960). Type: Malawi, Shire Highlands, *Buchanan* 484 (K, lecto. !)

Perennial herb 1–1·2 m. high, a few adventitious roots often present at base of stem; stem 4-angled, pilose (hairs ± 3 mm. long), cream or more often reddish. Leaf-lamina lanceolate to oblong-obovate, 6–11 cm. long, 1·7–5·7 cm. wide, apex acute, base rounded, laxly appressed pilose on both surfaces; midrib and 2 pairs of basal longitudinal nerves impressed above, subprominent beneath; petiole 2–7 mm. long. Inflorescence terminal dense subglobose cymes; flowers 5-merous. Calyx-tube (fig. 10/3, p. 31) cylindric-campanulate, 8–10 mm. long, 5·5 mm. in diameter, lower half of tube densely long-setose (setae bulbous-based, 3–4 mm. long), upper half covered with stellate-setose appendages, appendages ligulate, 2–2·5 mm. long, setae whitish or more often reddish and obscuring the surface; lobes ovate-lanceolate, 5–7 mm. long, 2–2·5 mm. wide, margin ciliate, apex acute, long stellate-setose; intersepalar appendages long stellate-setose. Petals broadly obovate, 14–18 mm. long, 8–14 mm. wide. Stamens 10, equal; anthers 8 mm. long; connective produced 1 mm.; bituberculate appendages 0·5 mm. long; filaments 8 mm. long. Fruiting calyx enlarging to 9–11 mm. long, 6 mm. in diameter; capsule setose at the apex, not exserted.

TANZANIA. Buha/Kigoma Districts: Gombe Stream Reserve, Mkenke valley, 25 Jan. 1964, *Pirozynski* 288!; Mbeya District: Mbozi, June 1967, *Knight* 204!; Songea District: near Mbamba Bay, 4 Apr. 1956, *Milne-Redhead & Taylor* 9516!

DISTR. **T**4, 7, 8; Zaire, Mozambique, Malawi, Zambia and Angola

HAB. Valley and upland grassland; 910–2100 m.

* Forma *osbeckioides* has not so far been recorded for the Flora area but is likely to be found since it occurs both to the north and the south.

SYN. *Osbeckia densiflora* Gilg in E.M. 2: 8 (1898), excl. specim. *Buchanan* 114; V.E. 3 (2): 746 (1921)

19. **D. princeps** (*Kunth*) *Triana* in Trans. Linn. Soc. 28: 57 (1872); Cogn. in A. & C. DC., Monogr. Phan. 7: 375 (1891); Taub. in P.O.A. C: 295 (1895), pro parte, excl. syn. *D. violacea* Gilg; Gilg in E.M. 2: 22 (1898); V.E. 3 (2): 753 (1921); T.T.C.L.: 310 (1949); A. & R. Fernandes in Mem. Soc. Brot. 11: 90 (1956) & in Bol. Soc. Brot., sér. 2, 34: 190 (1960); F.F.N.R.: 308 (1962); A. & R. Fernandes in C.F.A. 4: 154 (1970). Type: Mozambique, 1788, *da Silva* (P, holo.)*

Shrub or woody herb 1–3 m. high; stem 4–6-angled; branchlets very shortly setose. Leaf-lamina lanceolate to oblong-ovate, 4–16 cm. long, 1·5–5 cm. wide, apex acute, base rounded to subcordate, margin minutely crenulate with very shortly setulose teeth, often inconspicuously rugulose and sparsely and softly appressed setulose above or densely appressed pubescent with a mixture of single hairs and hispid appendages (× 20 lens required), softly tomentose to scabrid beneath; midrib and 2–3 pairs of basal longitudinal nerves impressed above, prominent beneath, reticulation subprominent beneath; petiole 1–5 cm. long. Inflorescence many-flowered congested or lax terminal panicles; flowers 5-merous; bracts caducous, ovate-orbicular or ovate, 9–15 mm. long, 5–12 mm. wide, strigillose. Calyx-tube (fig. 10/4, 5, p. 31) urceolate, 7–13 mm. long, 4–6 mm. in diameter, covered with long fascicled silky bristles intermixed with shorter bristles; short broad ligulate appendages present near the mouth of the tube; lobes oblong-ovate, oblong-lanceolate or obliquely oblong-lanceolate, 6–9 mm. long, 2–3 mm. wide, margin ciliate, pubescent; intersepalar appendages ligulate, 1·5 mm. long, apex long-setose. Petals violet to purple, broadly obovate, 22–30 mm. long, 18–23 mm. wide, margin minutely ciliate. Stamens 10; 5 long stamens with anthers 13 mm. long, connective produced 16–23 mm., slender, deeply bilobed appendage 3 mm. long, filament 15 mm. long; 5 short stamens with anthers 11 mm. long, connective produced 2–4 mm., bilobed appendage 2 mm. long, filament 15 mm. long. Fruiting calyx enlarging to 12 mm. long and 7 mm. in diameter; capsule setose at the apex, not exserted.

var. **princeps**; A. & R. Fernandes in An. Junta Invest. Ultramar 10: 39 (1955) & in Mem. Soc. Brot. 11: 91 (1956) & in C.F.A. 4: 155 (1970).

Fascicles of bristles on calyx-tube relatively dense, ± obscuring at least the lower half of the calyx-tube when young, ± 2·5 mm. long (fig. 10/5, p. 31). Inflorescence usually condensed.

TANZANIA. Njombe, by R. Ruhudji, 9 July 1956, *Milne-Redhead & Taylor* 11026!; Songea District: Matengo Hills, Lupemba Hill, 3 Mar. 1956, *Milne-Redhead & Taylor* 9009! & 9009A!; Masasi, Aug. 1965, *Beecher* 46!

DISTR. **T**7, 8; Cameroun, Zaire, Sudan (*fide* Cufodontis, E.P.A.: 630 (1959)), Ethiopia, Malawi, Mozambique, Zambia, Rhodesia, Angola, Swaziland and South Africa (Transvaal, Natal)

HAB. Upland rain-forest, riverine forest and secondary wooded grassland; 780–1860 m.

* Kunth in his original description of *Rhexia princeps* erroneously described the type locality as Brazil; the genus does not occur in S. America. Naudin, in Ann. Sci. Nat., sér. 3, 14: 54 (1850), decided that it could not be from Brazil but probably from the Cape Verde Islands. However it does not occur there either, and it was Triana, *loc. cit.* (1872), who identified the type with specimens collected from Natal. Triana also pointed out errors in both the drawings and the original description, which A. & R. Fernandes in An. Junta Invest. Ultramar 10: 44, 66 (1955) confirmed after examining the holotype. The erroneous type locality is explained by A. & R. Fernandes, *loc. cit.*: 46, 67 (1955), as being due to da Silva's Mozambique collection becoming mixed with a large collection from Brazil when sent from Lisbon to Paris.

SYN. *Rhexia princeps* Kunth, Voy. de Humboldt et Bonpland, 6, Monogr. Melast.: 122, t. 46 (1823)
Osbeckia ? princeps (Kunth) DC., Prodr. 3: 140 (1828); Naud. in Ann. Sci. Nat., sér. 3, 14: 54 (1850)
O. eximia Sond. in Linnaea 23: 48 (1850) & in Fl. Cap. 2: 518 (1862). Type: South Africa, Natal, Durban, *Gueinzius* 145 & 492 (both S, syn.)*
Dissotis eximia (Sond.) Hook. f. in F.T.A. 2: 454 (1871), excl. specimen *Grant*
D. verticillata De Wild. in B.J.B.B. 5: 79 (1915). Type: Zaire, Thsisinka, *Homblé* 1265 (BR, holo.)
D. bamendae Brenan & Keay in K.B. 7: 546 (1952). Type: Cameroun, Bamenda, *Keay* in *FHI* 28344 (K, holo. !)

var. **candolleana** (*Cogn.*) *A. & R. Fernandes* in Bol. Soc. Brot., sér. 2, 29: 56, t. 9, 10 (1955) & in Mem. Soc. Brot. 11: 40 (1956) & in C.F.A. 4: 155 (1970). Type: Angola, Malange, between Quitage and Condo, *Welwitsch* 404 (G, lecto., BM, BR, COI, K !, LISU, isolecto.)

Fascicles of bristles on calyx-tube less dense, not obscuring the calyx-tube, ± 1·5 mm. long (fig. 10/4, p. 31). Inflorescence usually lax.

TANZANIA. Rungwe District: Kyimbila, 23 Sept. 1910, *Stolz* 237 ! & Manow, 3 Feb. 1970, *Fuller* 28 !; Songea District: Matengo Highlands, Umgano, 12 May 1936, *Zerny* 666 !
DISTR. T7, 8; Cameroun (*fide* A. & R. Fernandes (1956), but no specimens seen), Zaire, Malawi, Mozambique, Zambia, Rhodesia, Angola, Swaziland and South Africa (Natal)
HAB. Valley and upland grassland; 1440–1950 m.

SYN. *D. candolleana* Cogn. in A. & C. DC., Monogr. Phan. 7: 375 (1891); Gilg in E.M. 2: 19 (1898); V.E. 3 (2): 751 (1921); T.T.C.L.: 309 (1949)
D. muenzneri Engl., V.E. 3 (2): 751 (1921); T.T.C.L.: 309 (1949). Types: Tanzania, Njombe District, Lumbila [Lumbira], collector not cited, presumably *Muenzner* (B, syn. †) & Rungwe District, Kyimbila, collector not cited but believed to be *Stolz* 237 (B, syn. †, BM, K, isosyn. !)

20. **D. denticulata** *A. & R. Fernandes* in Bol. Soc. Brot., sér. 2, 29: 57, t. 11 (1955) & 30: 177 (1956), descr. ampl., & in Mem. Soc. Brot. 11: 41, 93, t. 4 (1956); F.F.N.R.: 308, fig. 54 (1962). Type: Zambia, Shiwa Ngandu, *Greenway* 5765 (PRE, holo., EA, K, iso. !)

Shrub up to 2·5 m. high; branchlets setose; nodes thickened and long-setose. Leaf-lamina ovate-lanceolate, 2·5–11·5 cm. long, 1·5–5 cm. wide, apex acute, base subcordate, margin denticulate, bullate and densely setose above, softly and densely villose beneath; midrib and 2–3 pairs of basal longitudinal nerves impressed above, prominent beneath, reticulation prominent beneath; petiole 0·2–2·5 cm. long. Inflorescence of 3–4-flowered cymes aggregated into small terminal panicles; flowers 5-merous; bracts caducous, ovate-orbicular or ovate, up to 9 mm. long and 8 mm. wide. Calyx-tube (fig. 10/6, p. 31) cylindrical-campanulate, 8–10 mm. long, 6–7 mm. in diameter, densely covered with 2·5 mm. long capitate appendages bearing 2·5 mm. long setae; lobes obliquely oblong, 6–9 mm. long, 3·5–5 mm. wide, margin ciliate, otherwise appressed setulose; intersepalar appendages 3 mm. long, capitate, setose. Petals pinkish to purple, obcordiform, 25–28 mm. long, 17–20 mm. wide. Stamens 10, markedly unequal; 5 large stamens with anthers 12–15 mm. long, connective produced 13 mm., bilobed appendage ± 1 mm. long, filament 7–8 mm. long; 5 short stamens with anthers 9 mm. long, connective produced 3 mm., bituberculate appendage 0·5 mm. long, filament 10 mm. long. Fruiting calyx enlarging to 11 mm. long, 10 mm. in diameter; capsule with apical tuft of setae, not exserted.

TANZANIA. Iringa District: Mufindi, 8 Sept. 1932, *Geilinger* 2094 ! & Iheme, Oct. 1933, *Ward* 10 ! & Sao Hill, 16 Aug. 1949, *Greenway* 8429 !

* An unnumbered Gueinzius specimen in the Kew Herbarium is probably part of the type collection.

DISTR. **T7**; Zaire, Zambia and Angola
HAB. *Brachystegia* woodland, usually in damp places; 1500–1800 m.

21. **D. sessili-cordata** *Wickens* in K.B. 29: 146, fig. 4 (1974). Type: Tanzania, Mpanda District, Kabungu, *Semsei* 111 in *F.H.* 3542 (K, holo. !, EA, iso.)

Woody herb; stem 4-angled, shortly appressed pilose. Leaves sessile; lamina ovate, 1·5–5·5 cm. long, 0·7–2 cm. wide, apex acute, base cordate, appressed pilose on both surfaces; midrib and 1–2 pairs of longitudinal nerves impressed above, subprominent beneath. Inflorescences 1-flowered at the ends of leafy branches; flowers 5-merous, subtended by leaf-like persistent bracts. Flowers not seen. Fruiting calyx (fig. 11/1, p. 32) campanulate, 15 mm. long, 13 mm. in diameter, densely covered by setose scales, setae 1–7 mm. long; capsule not exserted, setose at the apex.

TANZANIA. Mpanda District: Kabungu, 3 Aug. 1948, *Semsei* 111 in *F.H.* 2542 !
DISTR. **T4**; not known elsewhere
HAB. Not known, said to be very common by roadsides; ± 2500 m.

NOTE. Further gatherings, especially of flowering material, of this very distinctive species are required.

22. **D. pachytricha** *R. E. Fries* in Wiss. Ergebn. Schwed. Rhod.-Kongo-Exped. 1911–1912, 1: 180, t. 13/7–11 (1914); V.E. 3 (2): 750 (1921); T.T.C.L. 310 (1949); A. & R. Fernandes in Mem. Soc. Brot. 11: 85 (1956) & in Bol. Soc. Brot., sér. 2, 34: 187 (1960); F.F.N.R.: 306 (1962). Types: Zaire, Lofonzo R., *Kassner* 2839 (B, syn. †, BM !, K !, isosyn.) & Tanzania, Ufipa District, Urungu, Kitungulu [Kistungulu], *Muenzner* 246 (B, syn. †) & Zambia, Malolo to Kitwe [Katwe], *R. E. Fries* 1203 (UPS, syn.)

Shrub or woody herb 1–2 m. high; branchlets weakly 4-angular, sparsely setose. Leaf-lamina lanceolate, lanceolate-elliptic or oblong-elliptic, 3–10 cm. long, 1–2·2 cm. wide, apex acute, base cuneate to subtruncate, margin subentire with a fringe of small setose hairs; glabrous or sparsely appressed-setose above, sparsely short white appressed setose beneath; midrib and 1 pair of longitudinal nerves impressed above, subprominent beneath; petiole 1–6 mm. long. Inflorescence terminal, with 1–few 5-merous flowers; bracts early caducous, ovate, 10 mm. long, 4 mm. wide, margin ciliate, midrib sparsely setose, otherwise glabrous. Calyx-tube (fig. 11/2–4, p. 32) subglobose to cylindric-campanulate, 9–18 mm. long, 5–9 mm. in diameter, densely covered with large reflexed scale-like appendages abruptly narrowed to 1 or several setae (setae obscuring the appendages in var. *orientalis*); lobes tardily deciduous, oblong to oblanceolate, 10–12 mm. long, 3·5–5 mm. wide, margin ciliate, otherwise glabrous or pubescent and bearing setose appendages down the centre; intersepalar appendages ligulate, setose. Petals reddish purple, obovate, 20–30 mm. long, 18–25 mm. wide. Stamens 10, markedly unequal; 5 large stamens with anthers 10–13 mm. long, connective produced ± 15–25 mm., bilobed filiform appendage 1–2 mm. long, filament 12–16 mm. long; 5 small stamens with anthers 9–12 mm. long, connective produced 2·5–3·5 mm., bilobed appendage 1 mm. long, filament 10–13 mm. long. Fruiting calyx enlarging to 10–20 mm. long and 8–12 mm. in diameter, appendages often somewhat accrescent; capsule setose at the apex, completely enclosed by the calyx.

KEY TO INFRASPECIFIC VARIANTS

Scale-like appendages on calyx-tube not obscured by hairs; calyx-lobes ciliate at the margins otherwise glabrous:

Calyx-tube ± 9 mm. long, 5 mm. in diameter, bearing ovate scale-like appendages with 1 or more long setae . . . . . . var. **pachytricha**
Calyx-tube ± 18 mm. long, 9 mm. in diameter, bearing transversely-elliptic scale-like appendages with several (3 or more) very short setae . . . . var. **grandisquamulosa**
Scale-like appendages on calyx-tube obscured by long hairs; calyx-lobes pubescent and medially setose . . . . var. **orientalis**

var. **pachytricha;** A. & R. Fernandes in Mem. Soc. Brot. 11: 85 (1956).

Leaf-lamina attenuate at base, glabrous above. Calyx-tube ± 9 mm. long, 5 mm. in diameter, densely clothed by ovate scale-like appendages terminating in 1 or more long setae. Fig. 11/4, p. 32.

TANZANIA. Mpanda District; Pasagulu, 6 Aug. 1959, *Harley* 9174! & Kungwe Mt., Kahoko, 22 July 1959, *Newbould & Harley* 4549!; Ufipa District: Mbala [Abercorn]–Sumbawanga road, 8 km. N. of border, 16 June 1960, *Leach & Brunton* 10055!
DISTR. **T4**; Zaire and Zambia
HAB. *Brachystegia* woodland and upland grassland; 1440–2100 m.

var. **grandisquamulosa** *Wickens* in K.B. 29: 144, fig. 2 (1974). Type: Tanzania, Kigoma District, S. of Uvinza, 1950, *Bullock* 3283 (K, holo.!)

Leaf-lamina subtruncate at base, sparsely appressed setose above. Calyx-tube 15–18 mm. long, 8–9 mm. in diameter, densely clothed by large transversely elliptic scale-like appendages bearing 3 or more very short (0·5 mm. long) setae. Fig. 11/2, p. 32.

TANZANIA. Buha District: Kalinzi C.M.S. Hospital, 22 Nov. 1962, *Verdcourt* 3396! & July 1963, *Bangham* H.264/63!; Kigoma District: 58 km. S. of Uvinza, 31 Aug. 1950, *Bullock* 3283!
DISTR. **T4**; not known elsewhere
HAB. Upland grassland; 1650–1710 m.

var. **orientalis** *A. & R. Fernandes* in Bol. Soc. Brot., sér. 2, 30: 176, t. 11 (1956) & in Mem. Soc. Brot. 11: 85 (1956) & in Bol. Soc. Brot., sér. 2, 34: 187 (1960). Type: Tanzania, Njombe District, Lupembe, Ruhudji R., *Schlieben* 510 (BR, holo.)

Leaf-lamina attenuate at the base, sparsely appressed setose above. Calyx-tube 12 mm. long, 9 mm. in diameter, densely clothed by ± ovate scale-like appendages bearing numerous setae (setae ± 5 mm. long, and obscuring the scales); calyx-lobes pubescent and bearing setose appendages down the centre. Fig. 11/3, p. 00.

TANZANIA. Njombe District: Lupembe, Ruhudji R., Apr. 1931, *Schlieben* 702!; Songea District: Mahenge, 1 July 1956, *Milne-Redhead & Taylor* 10740! & Maweso, July 1956, *Semsei* 2418!
DISTR. **T7, 8**; not known elsewhere
HAB. Upland grassland; ± 1250 m.

23. **D. simonis-jamesii** *Buscal. & Muschl.* in E.J. 49: 481 (1913); Gilg in E.J. 53: 371 (1915); A. & R. Fernandes in Bol. Soc. Brot., sér. 2, 46: 68 (1972). Type: Zambia, Bangweulu, *von Aosta* 927 (B, holo. †)

Woody herb (? suffrutex) 0·6–1·5 m. high; branches 4-angled; branchlets lightly to densely silvery pilose, when dense, indumentum appearing somewhat cobwebby. Leaf-lamina conduplicate and reflexed, lanceolate to lanceolate-ovate, 1–6 cm. long, 0·3–1·5 cm. wide, apex acute, base rounded, appressed silvery pilose or sometimes cobwebby on both surfaces; midrib and 2 pairs of longitudinal nerves lightly impressed above, faintly discernible beneath; petiole 0·5–2 mm. long or absent. Inflorescence of solitary flowers at the apex of branches; flowers 5-merous, subsessile; bracts reddish, broadly ovate, up to 15 mm. long and 15 mm. wide, shortly sericeous. Calyx-tube (fig. 11/5, p. 32) campanulate, 14 mm. long, 10 mm. in diameter, densely silvery sericeous, with simple hairs ± 7 mm. long on appressed linear appendages ± 1 mm. long; lobes tardily deciduous, oblong-ovate, 14 mm. long, 5 mm. wide, silvery sericeous except for the glabrous tip, margins ciliate; intersepalar appendages ovate, ± 2 mm. long, 1·5 mm. wide, coarsely pilose along the margins, the body sericeous. Petals obovate, 45–50 mm. long, 40 mm. wide. Stamens 10, markedly unequal; 5 large stamens with anthers 13–15 mm. long, free part of connective arcuate, 20–30 mm. long, bilobed appendage 1–3 mm. long, filament 15 mm. long; 5 short stamens with anthers 9–12 mm. long, connective produced 4 mm., bilobed appendage 0·5–1·5 mm. long, filament 15 mm. long. Fruiting calyx with setose apex of capsule slightly exserted.

TANZANIA. Ufipa District: between Lakes Tanganyika and Rukwa, *Nutt*! & 56 km. Mbala [Abercorn]–Sumbawanga, 3 Mar. 1951, *Bullock* 3740!
DISTR. **T4**; Zambia
HAB. Valley grassland; 1500–1800 m.

SYN. *D. degasparisiana* Buscal. & Muschl. in E.J. 49: 480 (1913); Gilg in E.J. 53: 371 (1915); A. & R. Fernandes in Mem. Soc. Brot. 11: 84 (1956) & in Kirkia 1: 74 (1961). Type: Zambia, Bangweulu, *von Aosta* 1135 (B, holo. †)

24. **D. trothae** *Gilg* in E.M. 2: 19, t. 2/B (1898); V.E. 3 (2): 751 (1921); A. & R. Fernandes in Mem. Soc. Brot. 11: 86 (1956) & in Bol. Soc. Brot., sér. 2, 34: 67, 188 (1960); F.F.N.R.: 310 (1962). Type: Burundi, Migera Hills, Umuesi, *von Trotha* 28 (B, holo.†)

Woody herb or shrub 0·3–3·5 m. high; branchlets 4-angular, densely squamulose-setose. Leaf-lamina ovate to ovate-lanceolate, 4–13(–24) cm. long, 2–5·5(–10·5) cm. wide, apex acute, base rounded to cordate, margin serrate-setose, bullate and densely long setose above, densely villous beneath; midrib and 2 pairs of longitudinal nerves impressed above, prominent beneath; reticulation prominent beneath; petiole 1–3·5(–6) cm. long. Inflorescence of crowded apical cymes; flowers 5-merous; pedicels ± 1 mm. long; bracts early deciduous, broadly ovate, ± 8 mm. long and 6 mm. wide, densely sericeous. Calyx-tube (fig. 11/6, p. 32) tubular-campanulate, 8–9 mm. long, 4 mm. in diameter, densely sericeous (hairs coarse, simple); lobes caducous, lanceolate-oblong, 6–7 mm. long, 2–3·5 mm. wide; intersepalar appendages oblong-ovate, 1–2 mm. long, sericeous. Petals broadly obovate, 25 mm. long, 20–25 mm. wide. Stamens 10, markedly unequal; 5 large stamens with anthers 10 mm. long, connective produced 12–15 mm., bilobed appendage 1 mm. long, filament 12 mm. long; 5 short stamens with anthers 9 mm. long, connective produced 3·5 mm. long, bilobed appendage ± 0·5 mm. long, filament 8 mm. long. Fruiting calyx 9 mm. long, 5 mm. in diameter, with setose apex of capsule not exserted.

UGANDA. Ankole District: Bushenyi, Apr. 1939, *Purseglove* 662! & Kitagata, 31 May 1970, *Lye* 5472!; Masaka District: Katera, 1 Oct. 1953, *Drummond & Hemsley* 4525!
TANZANIA. Bukoba District: Bugandika, 19 Apr. 1948, *Ford* 382!; Mpanda District: Inyonga, Msima R., 4 Sept. 1968, *Ruffo* 147!; Ufipa District: Kasanga–Sumbawanga, 21 June 1957, *Richards* 10180!
DISTR. **U**2, 4; **T4**; also in Zaire, Rwanda, Burundi and Zambia
HAB. Swamps and riverine; 1100–1650 m.

SYN. *D. mildbraedii* Gilg in Z.A.E.: 583 (1913). Type: Burundi, Rugege, *Mildbraed* 728 (B, holo. †)
[*D. caloneura* Engl., V.E. 3 (2): 749 (1921) pro specim. *Kassner* 2960 in herb. K, BM, pro majore parte, et P, *non* Engl. sensu stricto]
D. grandiceps Kraenzlin in Viert. Nat. Ges. Zürich 76: 149 (1931). Type: Zaire (Katanga), Kibandu, *Kassner* 3055 (Z, holo.)

25. **D. cryptantha** *Bak.* in K.B. 1894: 345 (1894); Gilg in E.M. 2: 17 (1898); V.E. 3 (2): 750 (1921); A. & R. Fernandes in Mem. Soc. Brot. 11: 83 (1956) & in Bol. Soc. Brot., sér. 2, 34: 186 (1960) & 43: 296 (1969) & 46: 68 (1972). Type: Malawi, *Buchanan* 625 (K, holo. !, BM, iso. !)

Shrub up to 2 m. high; branchlets densely appressed pilose. Leaf-lamina oblong to oblong-lanceolate, 1·8–3·8 cm. long, 0·7–1·3 cm. wide, apex acute, base rounded, appressed pilose on both surfaces; midrib and 2 pairs of longitudinal nerves impressed above, subprominent beneath; petiole 2–4 mm. long. Inflorescences 3-flowered cymes at apex of stem and solitary flowers at ends of branches; flowers 5-merous; pedicels very short, ± 1 mm. long; bracts absent (reduced leaves surround the flowers). Calyx-tube (fig. 12/1, p. 33) subglobose, 7–8 mm. long, 4·5–5·5 mm. in diameter, densely appressed sericeous, the indumentum almost concealing the small scale-like appendages to which the hairs are attached; lobes deciduous, lanceolate or linear-lanceolate, 8–10 mm. long, 2 mm. wide, acute, sericeous; intersepalar appendages persistent, oblong-ovate, 1·5 mm. long, 1 mm. wide, margin setose (setae 2·5–3·5 mm. long). Petals obovate, 27 mm. long, 17 mm. wide. Stamens 10, markedly unequal; 5 large stamens with anthers 9 mm. long, connective produced 9 mm., bilobed appendage 1·5 mm. long, filament 8 mm. long; 5 short stamens with anthers 8 mm. long, connective produced 2 mm., bilobed appendage 1 mm. long, filament 8 mm. long. Fruiting capsule enlarging to 11 mm. long and 8 mm. in diameter; capsule setose at the apex, not exserted.

TANZANIA. Ulanga District: Lupiro, near Merera, 29 July 1960, *Haerdi* 584/0!; Rungwe District: Rungwe foothills, 19 May 1913, *Stolz* 2000!; Songea District: W. of Songea, 13 June 1956, *Milne-Redhead & Taylor* 10729!
DISTR. T6–8; Zambia, Malawi and Mozambique
HAB. Valley grassland and upland forest; 990–2000 m.

SYN. *D. spectabilis* Gilg in E.J. 30: 366 (1901); V.E. 3 (2): 750 (1921); T.T.C.L.: 310 (1949); A. & R. Fernandes in Mem. Soc. Brot. 11: 32 (1956). Type: Tanzania, N. Mt. Rungwe, *Goetze* 1138 (B, holo. †, BM, iso. !)

26. **D. perkinsiae** *Gilg* in E.M. 2: 21, t. 3/0 (1898); V.E. 3 (2): 753, fig. 317/J (1921); F.W.T.A., ed. 2, 1: 258 (1954); A. & R. Fernandes in Bol. Soc. Brot., sér. 2, 34: 68 (1960). Types: Ghana, Odomi, *Kling* 138 (B, syn. †) & Togo, Missahohé, *Baumann* 320 (B, syn. †) & Bismarckbourg, *Buettner* 285 (B, syn. †, K, isosyn. !)

Woody herb or shrub 1–2·5 m. high; branchlets 4-angled, densely setose at first, becoming ± strigose. Leaf-lamina ovate, oblong-ovate or oblong-lanceolate, 3–11·5 cm. long, 1·2–5 cm. wide, apex obtuse to acute, base truncate to subcordate, densely strigose when young, becoming strigose with lower half of hairs adnate to the lamina; midrib and 2–3 pairs of longitudinal nerves impressed above, prominent beneath; petiole 2–10 mm. long. Inflorescence of terminal paniculate cymes; flowers 5-merous, subsessile; bracts caducous, ovate, 10 mm. long, 8 mm. wide, sericeous. Calyx-tube (fig. 12/2, p. 33) campanulate, 9–10 mm. long, 5–6 mm. in diameter, densely covered with oblong to obtriangular setose scales; lobes obliquely oblong, 10 mm. long, 3 mm. wide, sericeous; intersepalar appendages oblong-spathulate, 3 mm. long, 1·5 mm. wide, setose at the apex, alternating with oblong-ovate appendages 2 mm. long, 1 mm. wide. Petals magenta to purple, obovate, 30–40 mm. long, 20–30 mm. wide. Stamens 10, markedly unequal; 5 large stamens with anthers 13 mm. long, connective produced 17–24 mm., bilobed appendage ± 1·5 mm. long, filament ± 8–15 mm. long; 5 short stamens with anthers 8–11 mm. long, connective produced 5 mm., bilobed appendage 1 mm. long, filament ± 7 mm. long. Fruiting calyx enlarging to 13 mm. long, 8 mm. in diameter; setose apex of capsule shortly exserted.

UGANDA. W. Nile District: Mt. Otzi, Oct. 1959, *E. M. Scott* in *E.A.H.* 11799!; Toro District: Kyahara, 23 Jan. 1952, *A. M. S. Smith* 28!; Mengo District: Kasa Forest, 17 Nov. 1949, *Dawkins* 451!

DISTR. **U**1, 2, 4; Ghana, Togo, Nigeria, Cameroun, Zaire and Sudan

HAB. Wooded grassland and old fallow lands; 1200–1500 m.

SYN. [*D. eximia* sensu Hook. f. in F.T.A. 2: 454 (1871), pro parte, quoad specim. *Grant*, *non* Hook. f. sensu stricto]

D. schweinfurthii Gilg in E.M. 2: 21 (1898); V.E. 3 (2): 752 (1921); A. & R. Fernandes in Bol. Soc. Brot., sér. 2, 34: 67 (1960). Types: Zaire, Isingeria, 1870, *Schweinfurth* 3194 & Munsa, 1870, *Schweinfurth* 3445 (both B, syn. †)

? *D. violacea* Gilg in E.M. 2: 22 (1898); V.E. 3 (2): 753 (1921); A. & R. Fernandes in Mem. Soc. Brot. 11: 93 (1956). Types: Zaire, Abumbi-Ituri watershed, Sept. 1891, *Stuhlmann* 2674a & Walegga, Nov. 1891, *Stuhlmann* 2880 (both B, syn. †)

27. **D. speciosa** *Taub.* in P.O.A. C: 295 (1895); Gilg in E.M. 2: 18, t. 3/C (1899); V.E. 3 (2): 750 (1921). Type: Uganda, Mengo, *Stuhlmann* 1335 (B, holo. †)

Woody herb or shrub 0·5–2·5 m. high; stem 4-angled; branchlets densely appressed pubescent. Leaf-lamina lanceolate, lanceolate-ovate or narrowly oblong, 1·5–7 cm. long, 0·5–2 cm. wide, apex subacute, base rounded to truncate, white sericeous above at first, becoming appressed setulose, appressed pilose beneath; midrib and 2 pairs of longitudinal nerves impressed above, subprominent beneath; petiole 0·5–5 mm. long. Inflorescence terminal, flowers solitary, ± concealed at first by the leaves; flower 5-merous; bracts often reddish, tardily deciduous, ovate to ovate-lanceolate, 13–14 mm. long, 8–12 mm. wide, sericeous. Calyx-tube (fig. 12/3, p. 33) cup-shaped, 12–14 mm. long, 8–12 mm. in diameter, densely appressed sericeous-villous, hairs concealing the scale-like appendages; lobes caducous, lanceolate, 9–14 mm. long, 2–3 mm. wide, subacute, sericeous-villous; intersepalar appendages alternating with appendages opposite the sepals, semi-orbicular or semi-elliptic, 2–4 mm. long, pubescent, margins long setose (appendages often concealed by the indumentum of the calyx). Petals mauve to reddish purple, obovate, 35–40 mm. long, 22–27 mm. wide. Stamens 10; 5 large stamens with anthers 10–14 mm. long, connective produced 15–25 mm., 2 filiform appendages 2 mm. long, filament 15 mm. long; 5 short stamens with anthers 8–12 mm. long, connective produced 3–4 mm., 2 filiform appendages 1·5 mm. long, filament 13 mm. long. Fruiting calyx enlarging to 15 mm. long and 12 mm. in diameter; capsule setose at the apex, not exserted.

UGANDA. Mbale District: Busiu, July 1926, *Maitland*!; Masaka District: Katera, 4 Oct. 1953, *Drummond & Hemsley* 4614!; Mengo District: Kampala, Dec. 1921, *Ruck* in *Snowden* 2380!

KENYA. W. Suk District: 32 km. Kitale–Lodwar, 4 Oct. 1952, *Verdcourt* 734!; Trans-Nzoia District: Kitale Grassland Research Station, 2 Oct. 1959, *Verdcourt* 2467!; Masai District: Abossi, 6 July 1961, *Glover et al.* 2296!

TANZANIA. Ngara District: Bushubi, Keza, 20 May 1960, *Tanner* 4592!; Iringa District, Kilolo, 10 Feb. 1962, *Polhill & Paulo* 1433A!

DISTR. **U**3, 4; **K**2, 3, 5, 6; **T**1, 4, 7; Zaire, Sudan and Zambia

HAB. Valley grassland, swampy places; 900–2250 m.

SYN. *D. macrocarpa* Gilg in E.M. 2: 18 (1899); V.E. 3 (2): 750 (1921); T.T.C.L.: 309 (1949); F.P.S. 1: 194 (1950); A. & R. Fernandes in Mem. Soc. Brot. 11: 32, 83 (1956) & in Bol. Soc. Brot., sér. 2, 34: 187 (1960). Type: Uganda, Masaka District, Buddu, *Scott-Elliot* 7480 (B, holo. †, K, iso. !)

D. helenae Buscal. & Muschl. in E.J. 49: 479 (1913); Gilg in E.J. 53: 371 (1915). Type: Zambia, Lake Bangweulu, *von Aosta* 1346 (B, holo. †)

D. mirabilis Bullock in K.B. 1931: 99 (1931). Type: Uganda, Masaka District, Katera, Minziro Forest, *Brasnett* 22 (K, holo. !)

NOTE. Specimens with longer and denser indumentum and early caducous bracts from **T4** and **T7**, e.g. Ufipa District, Namanyere, 4 Apr. 1950, *Bullock* 2838 !, Mpanda District, Sobogo, 31 July 1958, *Newbould & Jefford* 1225 !, Sikoke, 29 Sept. 1958, *Newbould & Jefford* 2758 ! and Njombe, 29 Nov. 1931, *Lynes* S.16 ! represent an extreme form of this species that appears not to warrant taxonomic distinction.

28. **D. pterocaulos** *Wickens* in K.B. 29: 146, fig. 3 (1974). Type: Tanzania, Buha District: 48 km. Uvinza–Kasulu, *Verdcourt* 3427 (K, holo. !, EA, iso. !)

Woody herb 0·6–1 m. high; branchlets 4-winged (wings 1·5–2 mm. wide, narrowing at the nodes), densely woolly at first, becoming densely setose; nodes long setose, setae 7–10 mm. long. Leaf-lamina oblong-ovate, 4–6·5 cm. long, 0·7–1 cm. wide, apex acute, base shortly attenuate, woolly, becoming shortly strigose on both surfaces; midrib and 2 pairs of longitudinal nerves impressed above, prominent beneath; petiole 2–4 mm. long. Inflorescence terminal, 1-flowered or 3–5-flowered cymes; flowers 5-merous, subsessile; bracts absent. Calyx-tube (fig. 12/4, p. 33) campanulate, 8–9 mm. long, 5·5 mm. in diameter, densely sericeous, hairs simple, 3–7 mm. long; lobes caducous, linear-oblong, 15 mm. long, 2·5 mm. wide, sericeous; intersepalar appendages semicircular, 1 mm. long, setose (setae up to 8 mm. long). Petals obovate, 35 mm. long, 25 mm. wide. Stamens 10, markedly unequal; 5 large stamens with anthers 11 mm. long, connective produced 20 mm., bilobed appendage 2·5 mm. long, filament 15 mm. long; 5 short stamens with anthers 9 mm. long, connective 3 mm. long, bilobed appendage 1·5 mm. long, filament 18 mm. long. Fruiting capsule enlarging to 10 mm. long and 9 mm. in diameter; setose apex of capsule exserted.

TANZANIA. Buha District: 48 km. Uvinza–Kasulu, 23 Nov. 1962, *Verdcourt* 3427 !
DISTR. **T4**; not known elsewhere
HAB. Swamp; ± 1800 m.

29. **D. formosa** *A. & R. Fernandes* in Bol. Soc. Brot., sér. 2, 34: 65, t. 57 (1960) & 34: 187 (1960). Type: Tanzania, Songea District, Kwamponjore Valley, *Milne-Redhead & Taylor* 9924 (K, holo. !, COI, iso. !)

Woody herb 1·2 m. high; stem 4-angled; branchlets 4-winged (wings green, 1–2 mm. wide), nodes long-setose, otherwise sparsely appressed setose. Leaf-lamina narrowly ovate-elliptic, 1·5–4·5 cm. long, 0·6–1·4 cm. wide, apex acute, base rounded, appressed pilose on both surfaces at first, becoming glabrescent; midrib and 1 pair of lateral nerves impressed above, subprominent beneath; petiole 2 mm. long. Inflorescence terminal, of solitary subsessile flowers in the axils of the upper branches, forming leafy racemes; flowers (4–)5-merous; bracts caducous, broadly ovate, 15 mm. long, 12 mm. wide, apiculate, margin ciliate, surface pubescent. Calyx-tube (fig. 12/5, p. 33) broadly campanulate, 10–14 mm. long, 9–13 mm. in diameter, densely covered with obovate scales ± 0·7 mm. long and 0·5 mm. wide, the scales sericeous-setose, setae 1–3 mm. long and obscuring the scales; lobes caducous, linear-lanceolate, 13 mm. long, 4·5 mm. wide, densely sericeous; intersepalar appendages caducous, ligulate, ± 4 mm. long, 3 mm. wide, setose. Petals obovate, 35–40 mm. long, 25–30 mm. wide. Stamens 10 or 8; large stamens with anthers 11–13 mm. long, free part of connective arcuate, 20 mm. long, bilobed appendage 3–4 mm. long, filament 12–15 mm. long; short stamens with anthers 10–13 mm. long, connective produced 3 mm., bilobed appendage 3 mm. long, filament 11–14 mm. long. Fruiting calyx with setose apex of capsule exserted.

TANZANIA. Songea District: Kwamponjore Valley, 26 & 27 Apr. 1956, *Milne-Redhead & Taylor* 9924! & 9924A!
DISTR. T8; not known elsewhere
HAB. Valley grassland, marsh; 1000 m.

30. **D. alata** *A. & R. Fernandes* in Bol. Soc. Brot., sér. 2, 34: 60, t. 2 (1960). Type: Burundi, Kiofi, *Michel* 3745 (BR, holo., COI, YANG, iso.)

Woody herb up to 1·2 m. high; branchlets 4-winged (wings 1·5–2 mm. wide, narrowing at the nodes), appressed pubescent; nodes long-setose, setae 7–9 mm. long. Leaf-lamina linear-obovate, 2·5–7·5 cm. long, 0·4–0·8 cm. wide, apex acute, base attenuate, sericeous-villous on both surfaces when young, becoming appressed pilose; midrib and 2 longitudinal nerves impressed above, prominent beneath; petiole 2–5(–10) mm. long. Inflorescence terminal, 1-flowered or 2–3-flowered cymes; flowers 5-merous, subsessile. Calyx-tube (fig. 12/6, p. 33) campanulate, 7–8·5 mm. long, 7 mm. in diameter, densely appressed setose, 1–3 setae arising from minute tubercles, setae up to 1·5 mm. long; lobes caducous, linear-oblong, 16 mm. long, 3·5 mm. wide, densely appressed setose externally; intersepalar teeth semicircular, setose. Petals obovate, ± 24 mm. long, and 13 mm. wide. Stamens 10, markedly unequal; 5 large stamens with anthers 10 mm. long, connective produced 15 mm., with a bilobed appendage 1 mm. long, filament 12 mm. long; 5 short stamens with anthers 9 mm. long, connective produced 2 mm., bilobed appendage ± 0·8 mm. long, filament 12 mm. long. Ovary partially free. Fruiting calyx globose, enlarging to 9 mm. long, 8 mm. in diameter; capsule setose at the apex, setae exserted.

TANZANIA. Buha District: 115 km. Kibondo–Kasulu, 15 July 1960, *Verdcourt & Oteke* 2861!
DISTR. T4; Burundi
HAB. Valley grassland; 1350 m.

9. GRAVESIA

Naud. in Ann. Sci. Nat., sér. 3, 15: 333 (1851); Perrier de la Bâthie in Fl. Madag., 153 Melastom.: 73 (1951); A. & R. Fernandes in Bol. Soc. Brot., sér. 2, 30: 111 (1956)

Veprecella Naud. in Ann. Sci. Nat., sér. 3, 15: 315 (1851)
Phornothamnus Bak. in J.L.S. 21: 342 (1884)
Urotheca Gilg in E. & P., Pf., Nachtr. 3, 7: 263 (1897) & in E.M. 2: 28 (1898)
Petalonema Gilg in E. & P., Pf., Nachtr. 3, 7: 264 (1897) & in E.M. 2: 28 (1898), *non* Correns (1889), *nec* Schlechter (1915), *nec* A. Peter (1928)
Neopetalonema Brenan in Journ. Arn. Arb. 26: 213 (1945)

Shrubs or perennial herbs, erect, scandent or prostrate, in Madagascar sometimes acaulous. Leaves entire to serrate, with 3–9 longitudinal nerves usually but not always reaching to the base. Inflorescences terminal cymes, umbelliform, corymbose, paniculate or sometimes contracted to fascicles or solitary flowers. Flowers (4–)5-merous. Calyx turbinate, ± angled, smooth or ribbed, truncate or variously toothed or lobed. Petals ovate to obovate, white, pink or red. Stamens (8–)10, equal; anthers narrowly oblong to linear; connective not or scarcely extended below the anther, but usually produced into a posterior spur-like appendage, rarely with minute lateral tubercles. Ovary adherent to the calyx-tube to varying degrees, often largely by (8–)10 septa, depressed on the upper side with a surrounding crown of 5 firm accrescent scales. Capsule ± angular, crowned by the conspicuous accrescent often slightly woody plate- or wedge-like scales. Seeds narrowly oblong to obpyramidal.

About 110 species, 5 in tropical Africa, the rest in Madagascar.

This is a large and very diverse genus in Madagascar and is the only genus of *Sonerileae* recognized by Perrier de la Bâthie for that region. The African species were formerly ascribed to segregate genera, but there has been a progressive tendency to amalgamate them. *Calvoa* and the African acaulous genera do not fit conveniently within the natural subdivisions of the genus and have some exceptional features, particularly in the stamens, but there must be some doubt about their present status.

Anthers 5–8 mm. long; fruiting calyx 10–12 mm. in diameter; leaf-lamina sparsely setose when young, ± persistently ciliate, cuneate to cordate at the base, with (5–)7(–9) main nerves; erect woody herbs or shrubs . . . . 1. *G. pulchra*

Anthers 2–3 mm. long; fruiting calyx 7–9 mm. in diameter; leaf-lamina glabrous, base cuneate, with 5(–7) main nerves; semiprostrate woody herbs:

Pedicels 12–15 mm. long in flower, glandular-setose; branches slightly angled . . . . . 2. *G. hylophila*

Pedicels 5 mm. long in flower, glabrous; branches narrowly winged . . . . 3. *G. riparia*

1. **G. pulchra** (*Gilg*) *Wickens*, comb. nov. Type: Tanzania, Morogoro District, Uluguru Mts., *Stuhlmann* 8764, 8840 & 9289 (all B, syn. †)

Woody herb or shrub up to 3·5 m. high; branches sometimes sparsely to moderately glandular-setose when young, soon glabrous. Leaf-lamina ovate to elliptic, 2·5–20·5 cm. long, 2–10·5 cm. wide, acuminate, base cuneate or cordate, margins crenate, serrate or subserrate, ciliate, green above, purplish beneath, very sparsely setose above and on the nerves beneath when young, soon glabrous; midrib and (2–)3(–4) pairs of basal longitudinal nerves impressed above, prominent beneath, lateral nerves subprominent beneath; petiole 2–9 cm. long, tomentose at first, becoming glabrescent. Inflorescence (1–)3–15-flowered umbels; pedicels 10–15 mm. long, glandular-setose or glabrous. Calyx-tube 5–6 mm. long, 7 mm. in diameter, glabrous or glandular-pilose at first, becoming glabrous; lobes triangular, 1·5 mm. long, or linear-lanceolate, ± 6 mm. long. Petals 25–30 mm. long, 10 mm. wide, pink or purplish-red. Anthers 5–8 mm. long; filaments 11 mm. long. Calyx accrescent in fruit, increasing to 10 mm. long and 10–12 mm. in diameter, pedicels increasing to 30 mm. long, both becoming woody.

var. **pulchra**

Glandular setae absent. Leaf-lamina cordate at base, margins crenate. Inflorescence ± 15-flowered. Calyx-lobes linear-lanceolate.

TANZANIA. Morogoro District: NW. Uluguru Mts. 16 Oct. 1932, *Schlieben* 2811! & without precise locality, 16 Oct. 1932, *Wallace* 162! & Kitundu, 22 Nov. 1934, *E. M. Bruce* 172!

DISTR. T6; not known elsewhere

HAB. Upland rain-forest; 1200–1500 m.

SYN. *Petalonema pulchrum* Gilg in E. & P., Pf., Nachtr. 3, 7: 264 (1897) & in E.M. 2: 28, t. 4/B (1898); V.E. 3 (2): 755 (1921)

Neopetalonema pulchrum (Gilg) Brenan in Journ. Arn. Arb. 26: 213 (1945); T.T.C.L.: 313 (1949); A. & R. Fernandes in Mem. Soc. Brot. 11: 96, t. 7 (1956)

Gravesia pulchrum (Gilg) Jacques-Félix in B.J.B.B. 44: 169 (1974), *non rite publ.*

var. **glandulosa** (*A. & R. Fernandes*) *Wickens*, comb. nov. Type: Tanzania, Morogoro District, Uluguru Mts., Bondwa Peak, *Eggeling* 6454 (EA, holo., K. iso.!)

Young branches, petiole, pedicel and calyx-tube sparsely glandular-setose. Leaf-lamina usually cuneate at base, margins subserrate. Inflorescence (1–)3–6–flowered. Calyx-lobes triangular.

TANZANIA. Morogoro District: Nguru Mts., Ruhamba Peak, 2 Apr. 1953, *Drummond & Hemsley* 1984! & Uluguru Mts., Morningside, 5 Dec. 1934, *E. M. Bruce* 265! & 2 Oct. 1966, *B. J. Harris* 442! & Tanana, 24 Jan. 1935, *E. M. Bruce* 633!

DISTR. **T6**; not known elsewhere

HAB. Upland rain-forest; 1000–2000 m.

SYN. *Neopetalonema pulchrum* (Gilg) Brenan var. *glandulosum* A. & R. Fernandes in Bol. Soc. Brot., sér. 2, 43: 299, t. 9 (1969)
Gravesia pulchrum (Gilg) Jacques-Félix var *glandulosum* (A. & R. Fernandes) Wickens in K.B. 29: 592 (1974), *non rite publ.*

2. **G. hylophila** (*Gilg*) *A. & R. Fernandes* in Bol. Soc. Brot., sér. 2, 30: 115 t. 1 (1956) & in 43: 298, t. 8 (1969); Wickens in K.B. 29: 148 (1974). Types: Tanzania, Morogoro District, Nghwenn, *Stuhlmann* 8864 (B, syn. †, K, frag.!) & *Stuhlmann* 8761 (B, syn. †)

Semiprostrate woody herb 30–50 cm. high; young branches densely glandular-setose, the setae deciduous except for the persistent bases. Leaves anisophilous, opposite and decussate or sometimes ternate (*fide* Gilg (1898)); lamina oblong-elliptic or elliptic, 5–14·5 cm. long, 2–6 cm. wide, long-acuminate, base cuneate, glabrous; midrib and 2 pairs of longitudinal nerves impressed above, prominent beneath, lateral nerves subprominent beneath; petiole 1·5–6 cm. long. Inflorescence 2–15-flowered umbels; flowers 5-merous; peduncle ± 5 mm. long; pedicels 10–12 mm. long, glandular-setose. Calyx obconical, 4–5 mm. long, 5 mm. in diameter, glandular-setose; lobes shallowly triangular, ± 0·5 mm. high. Petals obovate, acute, 1·2–1·5 cm. long, 7–8 mm. wide, pink. Anthers ± 3 mm. long; posterior spur ± 1 mm. long; filaments 4 mm. long. Fruiting calyx accrescent and woody.

TANZANIA. Uluguru Mts., Nghwenn, Oct. 1894, *Stuhlmann* 8864! & Kinole, 21 Oct. 1932, *Schlieben* 2868a! & Tegetero, 20 Mar. 1953, *Drummond & Hemsley* 1726!

DISTR. **T6**; not known elsewhere

HAB. Upland rain-forest and evergreen bushland, sometimes in rocky places; 1000–1600 m.

SYN. *Urotheca hylophila* Gilg in E. & P., Pf., Nachtr. 3, 7: 264 (1897) & in E.M. 2: 28, t. 4/A (1898); V.E. 3 (2): 755, fig. 317/K (1921)
Orthogoneuron dasyanthum Gilg in E. & P., Pf., Nachtr. 3, 7: 267 (1897) & in E.M. 2: 36, t. 6/B (1898); V.E. 3 (2): 762 (1921); T.T.C.L.: 313 (1949). Type: Tanzania, Morogoro District, Uluguru Mts., Nghwenn, *Stuhlmann* 8861 & Tegetero, *Stuhlmann* 9035 (both B, syn. †)

3. **G. riparia** *A. & R. Fernandes* in Bol. Soc. Brot., sér. 2, 30: 115, t. 1 (1956) & in Mem. Soc. Brot. 11: 51 (1956). Type: Tanzania, Morogoro District, Uluguru Mts., *Schlieben* 3424 (PRE, holo., EA, LISC, iso.!)

Semi-prostrate perennial herb, up to 25 cm. high; branches 4-angled, lower ones bare, upper ones densely foliate and minutely red-lined. Leaf-lamina elliptic-lanceolate, up to 9·5 cm. long and 4 cm. wide, acuminate, base cuneate, glabrous; midrib and 2 pairs of longitudinal nerves conspicuous beneath. Inflorescences 2–5-flowered axillary and terminal subsessile umbelliform cymes; pedicel ± 5 mm. long. Calyx-tube obconical, 4 mm. long, 5 mm. in diameter, glabrous; lobes shortly ligulate, 2 mm. long. Petals trapeziform, long-acuminate, 18 mm. long, 7 mm. wide, red. Anthers 2 mm. long; posterior spur ± 1·5 mm. long; filaments 4 mm. long. Fruiting calyx accrescent and woody. Fig. 14.

TANZANIA. Uluguru Mts., 12 Feb. 1933, *Schlieben* 3424!

DISTR. **T6**; not known elsewhere

HAB. Upland rain-forest, streamside; 1550 m.

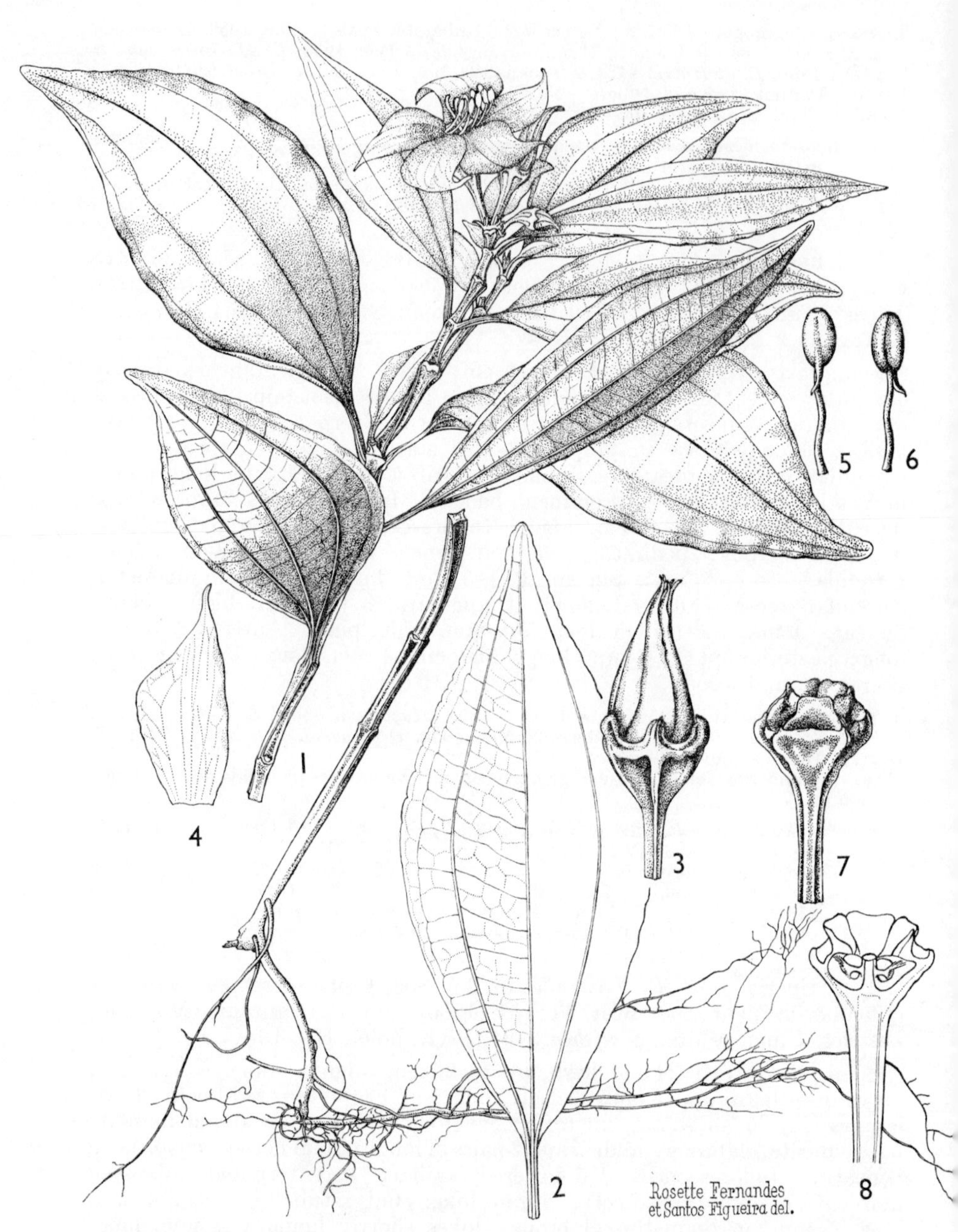

FIG. 14. *GRAVESIA RIPARIA*—**1,** flowering branch, × 1; **2,** leaf, × 1; **3,** flower bud, × 2; **4,** petal, × 2; **5, 6,** stamen, back and front views respectively, × 4; **7,** fruit, × 2; **8,** longitudinal section of capsule, × 2. All from *Schlieben* 3414. Reproduced by permission of the Editors of Boletim da Sociedade Broteriana.

10. CALVOA

Hook. f. in G.P. 1: 755 (1867) & in F.T.A. 2: 457 (1871); Gilg in E.M. 2: 31 (1898)

Succulent or semi-woody herbs. Leaves opposite, entire or dentate, long-petiolate. Inflorescence cymose; flowers secund, on short stout pedicels, 5-merous. Calyx-tube turbinate, turbinate-campanulate or tubular-campanulate, 5- or 10-ribbed or -angled; lobes small, deciduous. Petals obovate or oblong. Stamens 10, all equal or 5 smaller; anthers linear-oblong; connective little produced, with a small scale-like anterior appendage, the posterior appendage very small or lacking. Ovary adherent to the calyx-tube, 3–5-locular, concave on the upper side, with a surrounding crown of 3–5 firm scales. Capsule crowned by the accrescent woody wedge-like scales exserted well above the calyx. Seeds oblong or obovoid.

About 15 species in tropical Africa.

C. orientalis *Taub.* in P.O.A. C: 296 (1895); Gilg in E.M. 2: 32, t. 5/D (1898) & in V.E. 3 (2): 758, fig. 318/c (1921); A. & R. Fernandes in Mem. Soc. Brot. 11: 50 (1956). Types: Tanzania, Bukoba, *Stuhlmann* 994 & Lushoto District, Gonja Mt., *Holst* 4278 (both B, syn. †)

Semi-succulent or semi-woody herb up to 1 m. high, erect or semi-procumbent, sometimes epiphytic; branches scurfy and inconspicuously glandular-pubescent, glabrescent. Leaf-lamina elliptic, ovate or ovate-oblong, 2–7 cm. long, 1·5–4·5 cm. wide, apex acute, base cuneate to shortly attenuate, rarely rounded, margins slightly sinuate, serrate or dentate, ciliate; sparsely and inconspicuously glandular-puberulous on both surfaces, soon glabrous; midrib and 2 pairs of basal longitudinal nerves scarcely visible above, subprominent beneath, scurfy; petiole 1–5 cm. long. Inflorescence cymose; axis up to 11 cm. long; flowers 1–12, secund, on short stout pedicels ± 0·5 mm. long. Calyx turbinate, 3 mm. long, 4 mm. in diameter, 10-ribbed; lobes small, triangular-acuminate. Petals obliquely obovate, up to 10–15 mm. long, 8 mm. wide, pinkish-mauve. Stamens subequal; anthers 1·5–2 mm. long; anther connective produced ± 0·3–0·6 mm., anterior scale-lobe 0·5–1 mm. long, filaments 4 mm. long. Calyx accrescent in fruit to 4·5 mm. long, 5·5 mm. in diameter, woody, capsule with the crown of scales exserted beyond the calyx for ± 2·5 mm. Fig. 15.

UGANDA. Ankole District: Kalinzu Forest, June 1938, *Eggeling* 3648!; Kigezi District: Kayonza Forest Reserve, 4 Aug. 1960, *Paulo* 643!; Mengo District: Kyiwaga Forest, 7 Sept. 1949, *Dawkins* 366!

TANZANIA. Lushoto District: Amani, 6 Dec. 1928, *Greenway* 1030! & 2 Apr. 1950, *Verdcourt* 137!; Morogoro District: Nguru Mts., Mkobwe, 29 Mar. 1953, *Drummond & Hemsley* 1886!

DISTR. **U2**, 4; **T1**, 3, 6; Cameroun, Central African Republic and Zaire

HAB. Clearings and margins of rain-forest; 950–1800 m.

SYN. *C. sessiliflora* De Wild. & Th. Dur. in Ann. Mus. Congo, Bot., sér. 2, 1: 22 (1899). Type: Zaire, without precise locality, *Dewèvre* 1064a (BR, holo.)

NOTE. The plants from **T3** and **T6** are generally more robust, with better developed glandular indumentum, with the leaf margins conspicuously dentate, larger inflorescences and apparently larger petals, whereas the plants from Uganda are usually smaller, with serrate margins to the leaves and inflorescences, with up to 7 flowers per axis. There are intermediate forms, however, which suggest that any taxonomic distinction is undesirable, at least on present evidence.

FIG. 15. *CALVOA ORIENTALIS*—**1,** habit, × 1; **2,** part of leaf, lower surface, × 6; **3,** flower bud, × 4; **4,** flower, with petals removed, × 6; **5,** petal, × 6; **6, 7,** outer stamen, front and back views respectively, × 10; **8,** inner stamen, front view, × 10; **9,** detail of same, side view, to show appendages, × 20; **10,** longitudinal section of flower, × 8; **11,** fruit, × 4; **12,** seed, × 20. 1, 2, from *Faulkner* 1492; 3–11, from *Dawkins* 398; 12, from *Renvoize & Abdallah* 1596. Drawn by Mrs. M. E. Church.

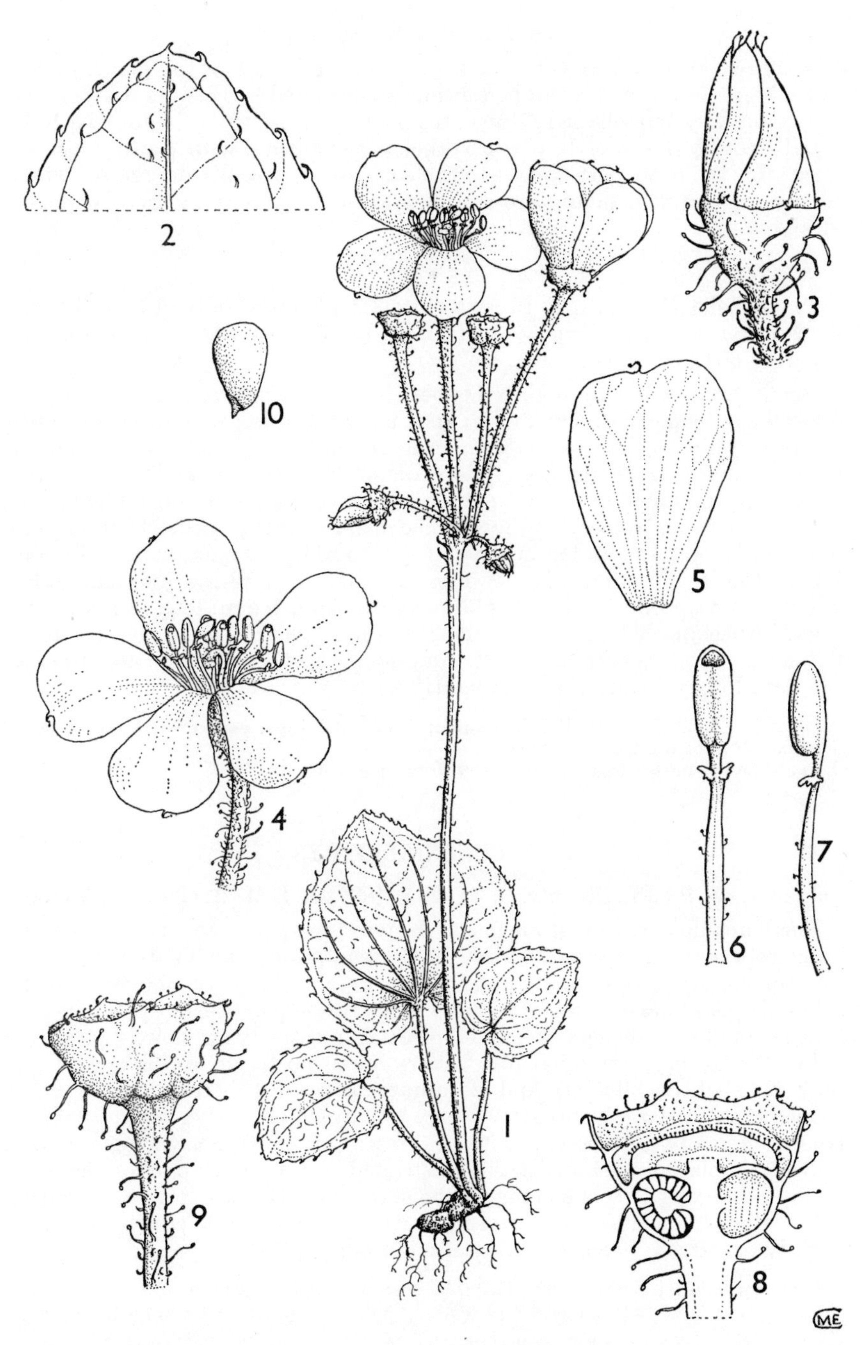

FIG. 16. *PRIMULARIA PULCHELLA*—**1,** habit, × 2; **2,** part of leaf, lower surface, × 4; **3,** flower bud, × 6; **4,** flower, × 3; **5,** petal, × 4; **6, 7,** stamen, front and side views respectively, × 12; **8,** longitudinal section of flower, × 12; **9,** fruit, × 8; **10,** seed, × 24. All from *Eggeling* 6041. Drawn by Mrs. M. E. Church.

11. PRIMULARIA

Brenan in K.B. 8: 88 (1953)

Acaulous herb. Leaves 1–few, arising from the base of the plant, shortly petiolate. Inflorescence umbelliform, few-flowered. Flowers 4–5-merous. Calyx broadly cup-shaped; lobes reduced to small teeth. Stamens 8–10, equal; connective shortly produced below the anther, with 2 anterior lobes. Ovary for the most part adnate to the calyx-tube, apically depressed, with a narrow glandular-ciliate crown, 4–5-locular. Seeds obovoid-cuneate, straight.

Monotypic.

P. pulchella *Brenan* in K.B. 8: 88 (1953); Jacques-Félix in Icon. Pl. Afr. 3, No. 69 (1955). Type: Tanzania, Lindi District, Rondo Plateau, *Eggeling* 6041 (K, holo. !, EA, iso.)

Small herb 6–9 cm. high. Leaves 1–3; lamina thin, broadly ovate, 0·7–2·8 cm. long, 0·5–2·4 cm. wide, apex usually obtuse, sometimes subacute, rarely acuminate, base cordate, margin subentire or minutely denticulate, shortly ciliate, laxly flexuous pilose on both surfaces; main nerves 7–9, palmate, prominent beneath; petiole 5–11 mm. long. Scape 4–6 cm. long, leafless, sparsely pilose. Inflorescence 2–6-flowered; pedicels 9–13 mm. long, glandular hairy and minutely scurfy; basal bracts lanceolate, ± 1·5 mm. long. Calyx 2·5 mm. long, 3 mm. in diameter, pilose and glandular-ciliate; lobes minute. Petals obliquely obovate, 7 mm. long, 6 mm. wide, pink. Anthers oblong, 1·5 mm. long, 0·5 mm. wide; connective produced 0·3 mm., lobes semi-ovate, 0·3 mm. long, toothed; filaments arcuate, 2·5 mm. long. Fruiting calyx slightly accrescent. Fig. 16.

TANZANIA. Lindi District, Rondo Plateau, below Mchinjiri, Feb. 1951, *Eggeling* 6041 !
DISTR. T8; not known elsewhere
HAB. Shaded mossy bank in evergreen forest; ± 1000 m.

12. CINCINNOBOTRYS

Gilg in E. & P., Pf., Nachtr. 3, 7: 265 (1897) & E.M. 2: 30, t. 6/A (1898)

Small acaulous rhizomatous herb. Leaves long-petiolate; lamina serrate, palmately nerved. Flowers in scorpioid cymes, 4-merous. Calyx campanulate-turbinate; lobes small, persistent. Petals obliquely oblong-obovate, minutely glandular-ciliate. Stamens 8, slightly unequal (the antipetalous ones a little smaller); anthers oblong-elliptic; connective very shortly produced below the anther, extended into a very small posterior appendage (which may be slightly bilobed) and 2 minute anterior tubercular appendages (which may be largely fused). Ovary adnate to the calyx for most of its length, apically depressed, with a crown of 4(–8) conspicuous scales, 4-locular; placentas short, sessile, restricted to the upper part of the axis; ovules numerous; style as long as the stamens; stigma capitate. Fruiting calyx slightly accrescent; capsule with the persistent crown of scales slightly exserted. Seeds obcuneate, minutely papillate.

Monotypic, but together with the other acaulous African genera of *Sonerileae* very similar to the large heteromorphic principally Madagascan genus *Gravesia* (see p. 60).

C. oreophila *Gilg* in E. & P., Pf., Nachtr. 3, 7: 265 (1897) & in E.M. 2: 30, t. 6/A (1898); V.E. 3 (2): 757, fig. 318/E (1921); A. & R. Fernandes in Bol. Soc. Brot., sér. 2, 43: 299, t. 10 (1969). Type: Tanzania, Uluguru Mts., Nghwenn, *Stuhlmann* 8804 (B, holo. †)

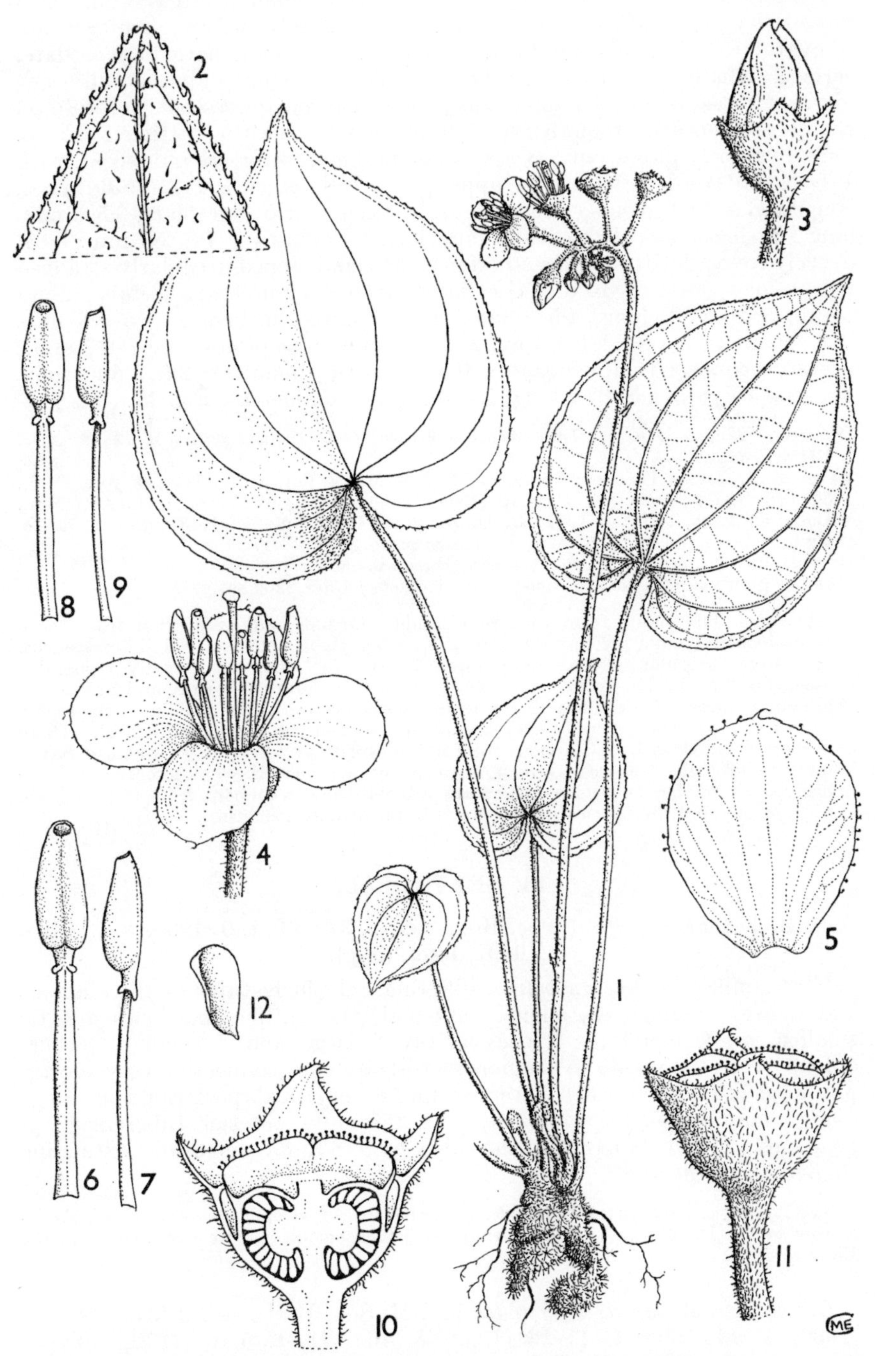

FIG. 17. *CINCINNOBOTRYS OREOPHILA*—**1**, habit, × 1; **2**, part of leaf, lower surface, × 6; **3**, flower bud, × 4; **4**, flower, × 4; **5**, petal, × 6; **6**, **7**, outer stamen, front and side views respectively, × 10; **8**, **9**, inner stamen, front and side views respectively, × 10; **10**, longitudinal section of flower, × 8; **11**, fruit, × 6; **12**, seed, × 24. All from *Greenway* 8658. Drawn by Mrs. M. E. Church.

Rhizomes slender or knobbly, with stellate trichomes, often swollen and more hairy just below the scape. Leaves 1–5, slightly fleshy; lamina ovate to ovate-orbicular, 4–15 cm. long, 3–10·5 cm. wide, acuminate, base cordate, serrate, ciliate, sometimes flushed purple beneath, scabrous-pubescent above with some longer hairs interspersed, scabrous-pubescent to setose on the nerves beneath; main nerves 7–9, prominent beneath; petiole 5–22 cm. long, retrorsely pubescent. Scape 8–30 cm. long, retrorsely pubescent, with 1–few small sterile bracts on the upper part. Inflorescence (1–)2–15-flowered, congested at first, usually becoming lax; bracts linear-lanceolate, 0·5–3 mm. long; pedicels 4–8(–15 in fruit) mm. long. Calyx-tube 1·5–2·5 mm. long, densely covered with simple and often some gland-tipped irregularly arranged hairs; lobes narrowly to broadly triangular, 0·5–3 mm. long. Petals pink or white, 5–10 mm. long, 4·5–8 mm. wide. Large anthers ± 1·5–1·8 mm. long, small anthers ± 1·1–1·4 mm. long; connective produced 0·1–0·3 mm.; posterior connective-appendage ± 0·5 mm. long; filaments 2·5–3 mm. long. Fruiting calyx 3–4·5 mm. long, 3·5–5 mm. in diameter. Fig. 17.

UGANDA. Kigezi District: Impenetrable Forest, Oct. 1940, *Eggeling* 4173! & Nov. 1948, *Eggeling* 5835!
TANZANIA. Kilosa District: Ukaguru Mts., Mamiwa Forest Reserve, Mnyera Peak, 30 July 1972, *Mabberley, Pócs & Alehe* 1280!; Morogoro District: Uluguru Mts., Lupanga Peak, 22 May 1933, *B. D. Burtt* 4717! & Aug. 1951, *Eggeling* 6272! & Ngula, above Bunduki, 23 Aug. 1951, *Greenway* 8658!
DISTR. **U2**; **T6**; eastern Zaire, Rwanda, Burundi
HAB. Upland rain-forest, in moist shaded places; 1500–2200 m.

VARIATION. The plants from western Uganda, Burundi and Kivu differ from those of eastern Tanzania by having only 1–3 rather than 6–15-flowered inflorescences, narrowly triangular aristate calyx-lobes 1·5–3 mm. long, not broadly triangular lobes 0·5–1·2 mm. long without a notably developed terminal bristle, and by lacking glandular hairs, which in the Tanzania specimens occur at least as cilia on the crown of ovary-scales and generally also among the hairs of the calyx and pedicel. There is clearly some incipient divergence (and interestingly *Gravesiella speciosa* has a mixture of these features), but it is uncertain whether these differences may be bridged by further collections (there are still very few specimens from the northern populations and the species may occur in intermediate localities).

13. GRAVESIELLA

A. & R. Fernandes in Bol. Soc. Brot., sér. 2, 34: 69, t. 9 (1960) & 43: 300 (1969), descr. ampl.

Very similar to *Cincinnobotrys*, differing only in features of the stamens and ovary. Stamens 8, slightly unequal (the antipetalous ones a little smaller) or 4, equal (or ? occasionally 2 large and 2 small); anthers oblong-ovate, attenuate to the narrow rostrate tip; connective very shortly produced below the anther, extended into a very small posterior appendage (which may be slightly bilobed) and 2 minute anterior tubercular appendages. Ovary adherent to the calyx by 4–8 septa; placentas extending the length of the axis.

Monotypic; in fruit virtually indistinguishable from *Cincinnobotrys* and *Haplophyllophorus seretii* (De Wild.) A. & R. Fernandes (which occurs just west and south of the Flora area).

G. speciosa *A. & R. Fernandes* in Bol. Soc. Brot., sér. 2, 34: 69, t. 9 (1960) & 43: 300, t. 11–13 (1969) & 46: 71, t. 5, 6 (1972). Type: Tanzania, Buha District, Kalinzi, *Procter* 612 (EA, holo.!)

Rhizomes with stellate trichomes, swollen and densely hairy just below the scape. Leaves 1–3; lamina ovate-orbicular, 4·5–22 cm. long, 4·5–25 cm. wide, acuminate, base cordate, serrate, finely setose and sometimes with small glandular hairs above, with rather long hairs on the nerves beneath; main nerves 7–13; petiole 3–40 cm. long, finely setose with generally a mixture of simple and gland-tipped mostly retrorse hairs. Scape up to 18–45 cm. long, with hairs similar to the petiole and with sterile bracts on the upper part. Cymes 3–26-flowered; bracts linear-lanceolate to narrowly elliptic, ± 2–3·5 mm. long; pedicels up to 8(–20 in fruit) mm. long. Calyx-tube ± 3 mm. long and wide, ± densely covered with simple or simple and gland-tipped hairs; lobes narrowly triangular-acuminate, 3–4 mm. long. Petals pink or white, broadly elliptic, ± 9 mm. long, 6 mm. wide. Large anthers ± 4 mm. long, small anthers (if present) ± 2·3 mm. long; connective produced ± 1 mm. and up to 1 mm. thick with the appendages; filaments 5·5–6 mm. long. Fruiting calyx 3–4 mm. long, 3–4 mm. in diameter; capsular crown of scales slightly exserted from the calyx, glandular-ciliate.

var. **speciosa**

Leaf-lamina up to 13·5 cm. long and 12·5 cm. wide; petiole up to 32 cm. long. Scape up to 30 cm. long. Inflorescence up to 15-flowered.

TANZANIA. Buha District: Kalinzi, Dec. 1956, *Procter* 612! & Mkenke stream valley, 28 Mar. 1964, *Pirozynski* 624!
DISTR. T4; Rwanda, Burundi
HAB. Rain-forest, in moist shaded places, sometimes epiphytic; 800–1500 m.

NOTE. Var. *grandifolia* A. & R. Fernandes (*loc. cit.*, 1972) is altogether more robust and known from a single locality in Burundi.

14. DICELLANDRA

Hook. f. in G.P. 1: 757 (1867) & in F.T.A. 2: 459 (1871); Stapf in J.L.S. 34: 490 (1900); Jacques-Félix in Adansonia, sér. 2, 14: 77 (1974)

Herb or undershrub, often rooting along the stems. Inflorescence a small terminal pyramidal panicle of cymes. Flowers (4–)5-merous. Calyx-tube campanulate-turbinate, sometimes slightly angular; lobes broadly triangular or rounded, short, with a slight to marked median crest. Petals lanceolate to oblong-oblanceolate. Stamens (8–)10, unequal or less often subequal; anthers linear-subulate, the larger ones sometimes curved and attenuate at the base; connective-appendages produced from the base of the anthers into 2 anterior tubercles or (on long stamens) upcurved spurs and 1 pendent posterior spur. Ovary partly adnate to the calyx, in the upper part or for the most part only by (8–)10 septa, the apex truncate with a membranous or little thickened crown, (4–)5-locular. Fruit a capsule within the accrescent calyx, the crown persisting or disappearing, but not accrescent. Seeds semi-elliptic or obovoid, attenuate at the base, and with the raphe produced into a lateral spur near the apex.

Three species in tropical Africa

The genus has generally been placed in the tribe *Dissochaeteae*, but as circumscribed by Jacques-Félix, *loc. cit.* (1974), it certainly belongs to the *Sonerileae*, though anomalous in the poor development of the crown on the ovary.

Dicellandra may be confused with *Phaeoneuron* Gilg, which has not yet been recorded for the Flora area, although known from Zaire. *Phaeoneuron* can be readily distinguished by the ovary not being adnate to the calyx-walls and by the wedge-shaped seeds without a spur.

D. barteri *Hook. f.* in G.P. 1: 757 (1867) & in F.T.A. 2: 459 (1871); Triana in Trans. Linn. Soc. 28: 81, t. 7/85 (1871); Cogn. in A. & C. DC., Monogr. Phan. 7: 546 (1891); Gilg in E.M. 2: 33 (1898); Stapf in J.L.S.

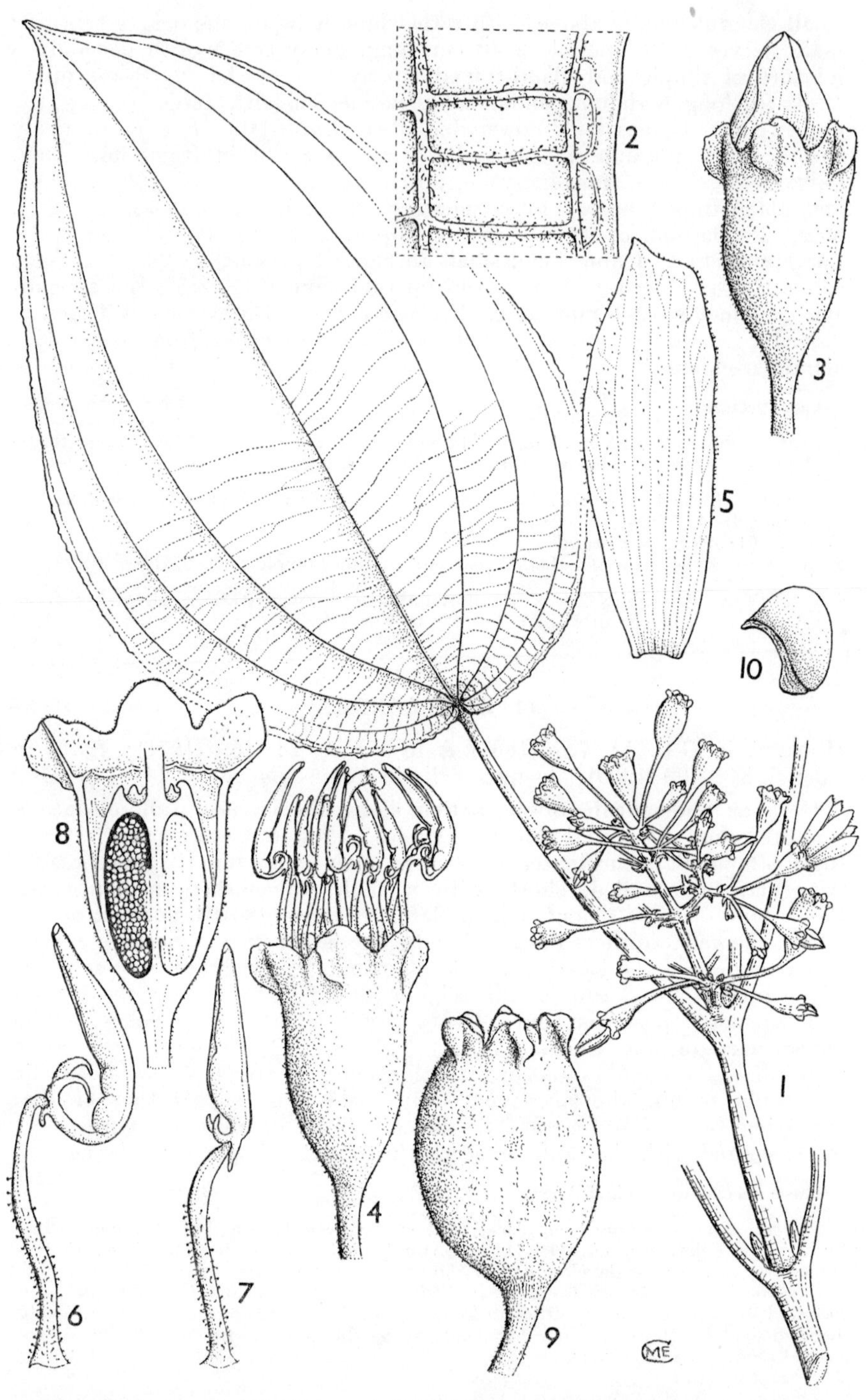

FIG. 18. *DICELLANDRA BARTERI* var. *MAGNIFICA*—**1,** fertile branchlet, × $\frac{2}{3}$; **2,** detail of leaf, lower surface, × 6; **3,** flower bud, × 3; **4,** flower, with petals removed, × 3; **5,** petal, × 3; **6,** outer stamen, × 6; **7,** inner stamen, × 6; **8,** longitudinal section of flower, × 4; **9,** fruit, × 3; **10,** seed, × 24. 1, 2, 9, 10, from *Purseglove* 3593; 3–8, from *Purseglove* 2634. Drawn by Mrs. M. E. Church.

34: 491, t. 19/1–10 (1900); Gilg in V.E. 3 (2): 760 (1921); F.W.T.A., ed. 2, 1: 252 (1954); Jacques-Félix in Icon. Pl. Afr. 3, No. 59 (1955); A. & R. Fernandes in Mem. Soc. Brot. 11: 47 (1956); Jacques-Félix in Adansonia, sér. 2, 14: 86, t. 4–7 (1974). Types: Nigeria, R. Nun, *Barter* 2113 & Fernando Po, *Mann* 3 (both K, syn.!)

Herb or undershrub 1–2 m. high, sometimes scandent or an epiphyte, rooting along the stem. Leaf-lamina ovate or ovate-orbicular, 5·5–35 cm. long, 3–21 cm. wide, apex acute, base rounded or cordate, margin scarcely dentate; midrib and 2–3 pairs of basal longitudinal nerves impressed above, prominent beneath, transverse nerves prominent beneath, nerves furfuraceous beneath when young; petiole 1–21 cm. long. Inflorescence ± 30–40-flowered. Calyx-tube 8 mm. long, 4 mm. in diameter, furfuraceous to nearly glabrous; lobes 1–2 mm. high. Petals 13–18 mm. long, 4–5 mm. wide, white or white flushed with pink. Large anthers 5–10 mm. long, anterior spur 0·7–1 mm. long, posterior spur ± 0·2 mm. long, filament up to 8 mm. long; short anthers 3–7 mm. long, anterior spur 0·2–0·5 mm. long, posterior spur 1 mm. long and parallel to the filament, filament up to 5·5 mm. long. Calyx-tube accrescent in fruit, enlarging to 13 mm. long, 8 mm. in diameter, enveloping the capsule.

var. **magnifica** (*Mildbr.*) *Jacques-Félix* in Adansonia, sér. 2, 14: 92, t. 6 (1974). Type: Zaire, Mawambi–Awakubi, near Abaranga, *Mildbraed* 3131 (B, holo. †)

Leaf-lamina membranous, ovate, at least 1·5 times as long as broad, base cordate; transverse nerves parallel. Cymes pedunculate, scorpioid; pedicels up to 15–20 mm. long. Fig. 18.

UGANDA. Kigezi District: Impenetrable forest, Mar. 1946, *Purseglove* 1963! & Ishasha Gorge, Apr. 1948, Aug. 1949 & Mar. 1951, *Purseglove* 2634!, 3058! & 3593!
DISTR. **U2**; Congo (Brazzaville) and Zaire
HAB. Rain-forest; 1200–1500 m.

SYN. *D. magnifica* Gilg in Z.A.E.: 586 (1913)

DISTR. (of species as a whole) **U2**; Guinée to Cameroun, Zaire and Angola

15. **MEDINILLA**

Gaud. in Freycinet, Voy. Bot.: 484 (1826)

Epiphytic or terrestrial shrubs, sometimes scandent. Leaves opposite to verticillate, or rarely alternate, equal or anisophyllous; lamina with 1–many pairs of longitudinal nerves, rarely without. Inflorescences terminal and/or axillary pedunculate panicles or cymes or rarely fascicles; bracts and bracteoles usually present. Flowers 4–5–6-merous. Calyx campanulate or cylindrical and widened at the top, thin or fleshy, entire, minutely dentate or with irregular small lobes. Petals equal, rarely unequal, usually red or white. Stamens 8–12, equal, subequal or unequal; anthers subequal or unequal, opening by 1(–2) pores; connective not or hardly produced at the base, posteriorly terminating in a subulate or subulate-clavate spur, anteriorly terminating in 2 lobes or 2 linear appendages or exappendiculate. Ovary adnate to the calyx, usually with 8, 10 or 12 septa above the middle (in Africa wholly adnate for most of its length), 4–5–6-locular; style filiform; stigma inconspicuous or rarely capitate. Fruit a berry enclosed within the persistent calyx. Seeds many, semi-ovate, smooth or punctate.

About 400 species, widespread through Madagascar, tropical Asia and the Pacific, with two species in Africa.

M. magnifica Lindl., a native of the Philippines, is cultivated in Tanzania at Amani, e.g. *Greenway* 3675!, see T.T.C.L.: 311 (1949).

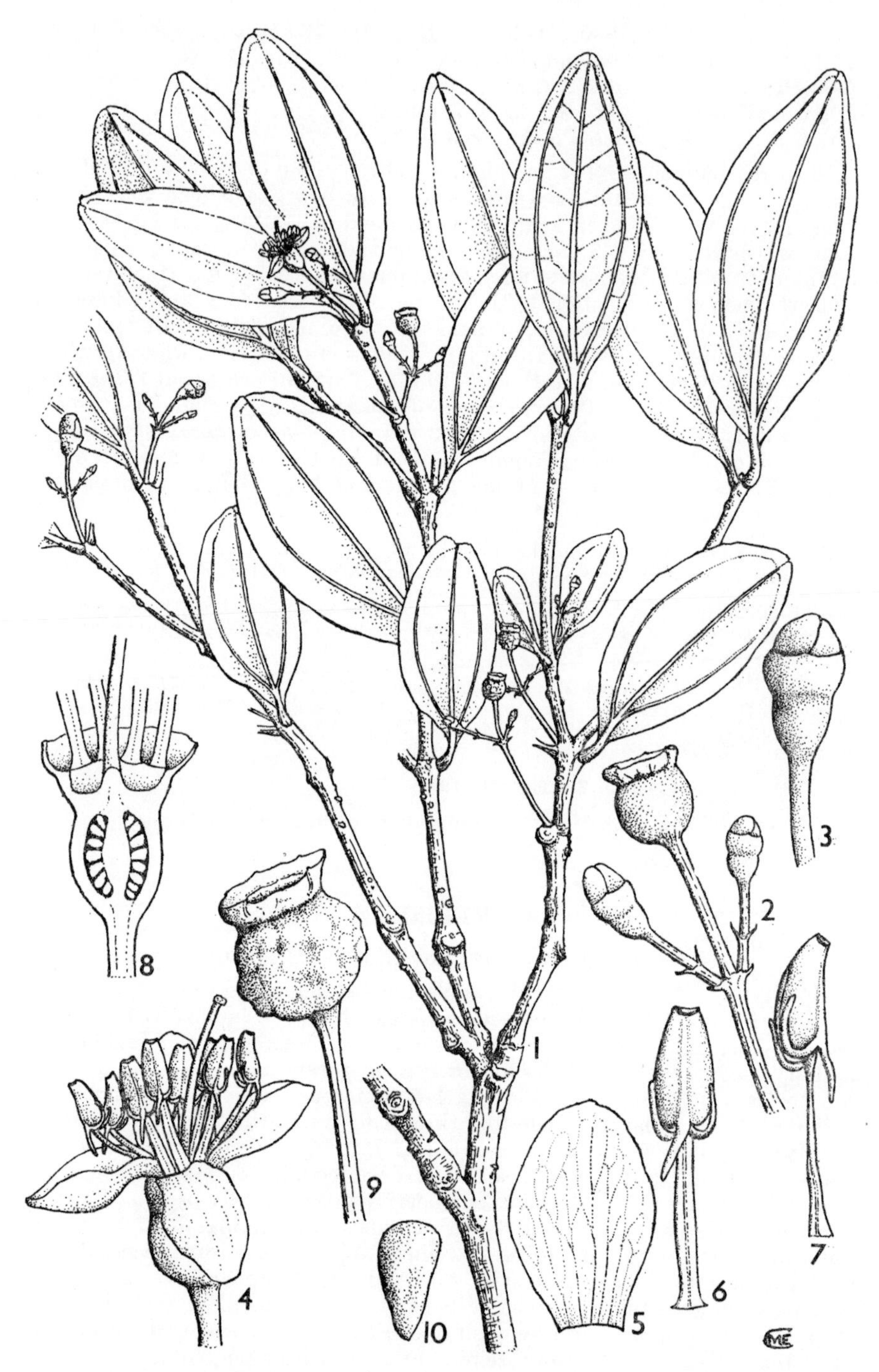

FIG. 19. *MEDINILLA ENGLERI*—**1**, habit, × 1; **2**, inflorescence, × 4; **3**, flower bud, × 6; **4**, flower, × 4; **5**, petal, × 6; **6**, **7**, stamen, back and side views respectively, × 8; **8**, longitudinal section of flower (petals removed), × 4; **9**, fruit, × 4; **10**, seed, × 16. All from *E. M. Bruce* 335. Drawn by Mrs. M. E. Church.

Leaf-lamina elliptic to ovate-elliptic, 1·5–2 times as long as broad, apex emarginate, base cuneate . . . 1. *M. engleri*
Leaf-lamina lanceolate-elliptic to elliptic, (2–)3–4 times as long as broad, acuminate, base cordate . . . 2. *M. mannii*

1. **M. engleri** *Gilg* in E.J. 34: 99 (1904); V.E. 3 (2): 760, fig. 319 (1921); T.T.C.L.: 311 (1949); A. & R. Fernandes in Mem. Soc. Brot. 11: 49 (1956). Types: Tanzania, Lushoto District, Amani, *Engler* 570 (B, syn.†) & *Warnecke* in *Herb. Amani* 387 (B, syn. †, EA, isosyn.!)

Epiphytic shrub up to 30–50 cm. high, often pendulous; roots thick, fleshy. Leaf-lamina elliptic to ovate-elliptic, 2–4·5(–6·5) cm. long, 1·2–2·8-(–3·8) cm. wide, apex emarginate, base cuneate, both surfaces glabrous; midrib and 1–2 pairs of longitudinal nerves slightly raised above, prominent beneath; petiole 1–3 mm. long. Inflorescence 1–5-flowered axillary cymes; peduncle 2–10 mm. long; pedicels 3–15 mm. long. Flowers 4-merous. Calyx campanulate, 3 mm. long, 2 mm. in diameter, glabrous apart from minute glandular cilia; lobes little developed, ± 0·5 mm. high. Petals obovate, 4 mm. long, 3 mm. wide, waxy red. Anthers equal, 1·2–1·5 mm. long; connective produced immediately below the anther into 2 upcurved anterior spurs 0·7–0·8 mm. long and 1 pendent posterior spur of similar length; filaments 3 mm. long. Fruit globose, ± 4–5 mm. in diameter. Fig. 19.

Tanzania. Lushoto District: Amani, 21 Dec. 1928, *Greenway* 1062! & Bumbuli Mission, 10 May 1953, *Drummond & Hemsley* 2481!; Morogoro District: Lupanga Peak, 8 Nov. 1932, *Schlieben* 2930!
Distr. T3, 6; not known elsewhere
Hab. Rain-forest; 900–1850 m.

2. **M. mannii** *Hook. f.* in F.T.A. 2: 460 (1871); Triana in Trans. Linn. Soc. 28: 87 (1872); Gilg in E.M. 2: 34 (1897); V.E. 3 (2): 760 (1921); F.W.T.A., ed. 2, 1: 251 (1954). Type: Fernando Po, *Mann* 302 (K, holo.!)

Epiphytic glabrous shrub up to 1 m. high; roots thick, fleshy. Leaf-lamina lanceolate-elliptic to elliptic, 3·5–12·5 cm. long, 1·5–4 cm. wide, acuminate, narrowed to the shortly cordate or almost auriculate base; midrib and 1–3 pairs of longitudinal nerves slightly raised above, sub-prominent beneath; petiole 2–4 mm. long. Inflorescence 1–20-flowered from axils of current and fallen leaves. Flowers 4-merous. Calyx campanulate, 2–2·5 mm. long, 4·5 mm. in diameter; lobes little developed, ± 0·5 mm. high. Petals oblanceolate, 6 mm. long, 3 mm. wide, white or red. Anthers 1·5–2·2 mm. long; connective produced immediately below the anther into 2 erect anterior spurs and 1 pendent posterior spur; filament 3–4 mm. long. Fruit globose, 6 mm. in diameter.

Uganda. Kigezi District: Impenetrable Forest, Apr. 1948, *Purseglove* 2676!; Masaka District: Sese Is., Towa Forest, 30 June 1935, *A. S. Thomas* 1358! & Bugala I., Kalangala, 24 Feb. 1945, *Greenway & Thomas* 7159!
Distr. U2, 4; also in Liberia, Ivory Coast, Ghana, Fernando Po and Zaire
Hab. Rain-forest; 1110–1500 m.

Syn. *M. afromontana* Lebrun & Taton in B.J.B.B. 18: 290 (1947); Jacques-Félix in Icon. Pl. Afr. 3, No. 62 (1955); A. & R. Fernandes in Mem. Soc. Brot. 11: 49 (1956). Type: Zaire, Kikomero, *Lebrun* 8453 (BR, holo., K, iso.!)

16. CLIDEMIA

D. Don in Mem. Wern. Nat. Hist. Soc. 4: 306 (1823); Gleason in Brittonia 3: 97–140 (1939)

Shrubs, erect or less often trailing to climbing and sometimes rooting between the nodes. Leaves sometimes anisophyllous, ± 3–9-nerved, the inner nerves sometimes meeting the midrib above the base of the lamina. Inflorescence of axillary fascicles, panicles or spikes, sometimes also terminal. Flowers 4–5–7-merous. Calyx-tube campanulate; lobes usually small, with a relatively well-developed dorsal appendage. Petals oblong to obovate, blunt or retuse. Stamens equal; anthers linear-subulate; connective not or very shortly produced below the anther, then sometimes lobed or extended into a small posterior appendage. Ovary adhering to the calyx in varying degrees, with a rim or crown of scales around the style-base, 3–5-locular. Fruit a berry within the accrescent calyx. Seeds very small, ± obovoid.

About 150 species, widespread in tropical America, one species naturalized in tropical Africa and elsewhere.

C. hirta (*L.*) *D. Don* in Mem. Wern. Nat. Hist. Soc. 4: 309 (1823), pro parte, excl. syn. *Melastoma hirta* sensu Miller (1768); D.C., Prodr. 3: 157 (1828); Triana in Trans. Linn. Soc. 28: 135 (1872); Cogn. in A. & C. DC., Monogr. Phan. 7: 986 (1891); Gleason in Brittonia 3: 108 (1939); A. & R. Fernandes in Mem. Soc. Brot. 11: 52 (1956) & Bol. Soc. Brot., sér. 2, 34: 193 (1960). Type: Central America, precise locality and collector not known (LINN, holo., IDC microfiche!)

Terrestrial shrub or facultative epiphyte 1–3 m. high; stem hirsute. Leaves slightly anisophyllous; lamina ovate to broadly ovate, up to 15 cm. long and 8 cm. wide, abruptly acuminate, base rounded to subcordate, margin crenate and long-ciliate, sparsely hirsute above and on the venation beneath, 5-nerved; nerves impressed above, prominent beneath, parallel venation prominent beneath; petiole up to 2·5 cm. long, hirsute. Inflorescence axillary, 5–12-flowered; flowers 5-merous. Calyx-tube campanulate, 4·5–5 mm. long, hirsute; lobes obscure, rounded-triangular, ± 0·5–1 mm. long; calyx-lobe appendages subulate, ± 4 mm. long, hirsute. Petals obovate-oblong, 8–11 mm. long, white to pale mauve. Stamens equal; anthers subulate, 4·5 mm. long; anther connective produced ± 0·2–0·3 mm. with a minute posterior lobe. Ovary adnate to the calyx-tube for one third of its length, tipped with a terete, glabrous neck; style ± 8 mm. long. Fig. 20.

TANZANIA. Lushoto District: Amani, 26 July 1921, *Soleman* in *Herb. Amani* 6093! & Kwamkoro Forest Reserve, 17 Nov. 1966, *Semsei* 4136!; Tanga District: Muheza, 2 Oct. 1955, *Tanner* 2257!

DISTR. T3; native of tropical America, also introduced and naturalized in the Ascension I., Madagascar, Indo-Malaysia and Polynesia

HAB. Weed of wayside and open spaces in lowland rain-forest, locally dominant; 210–1000 m.

SYN. *Melastoma hirta* L., Sp. Pl.: 390 (1753)

NOTE. See A. & R. Fernandes in Mem. Soc. Brot. 11: 52 (1956) for further synonymy.

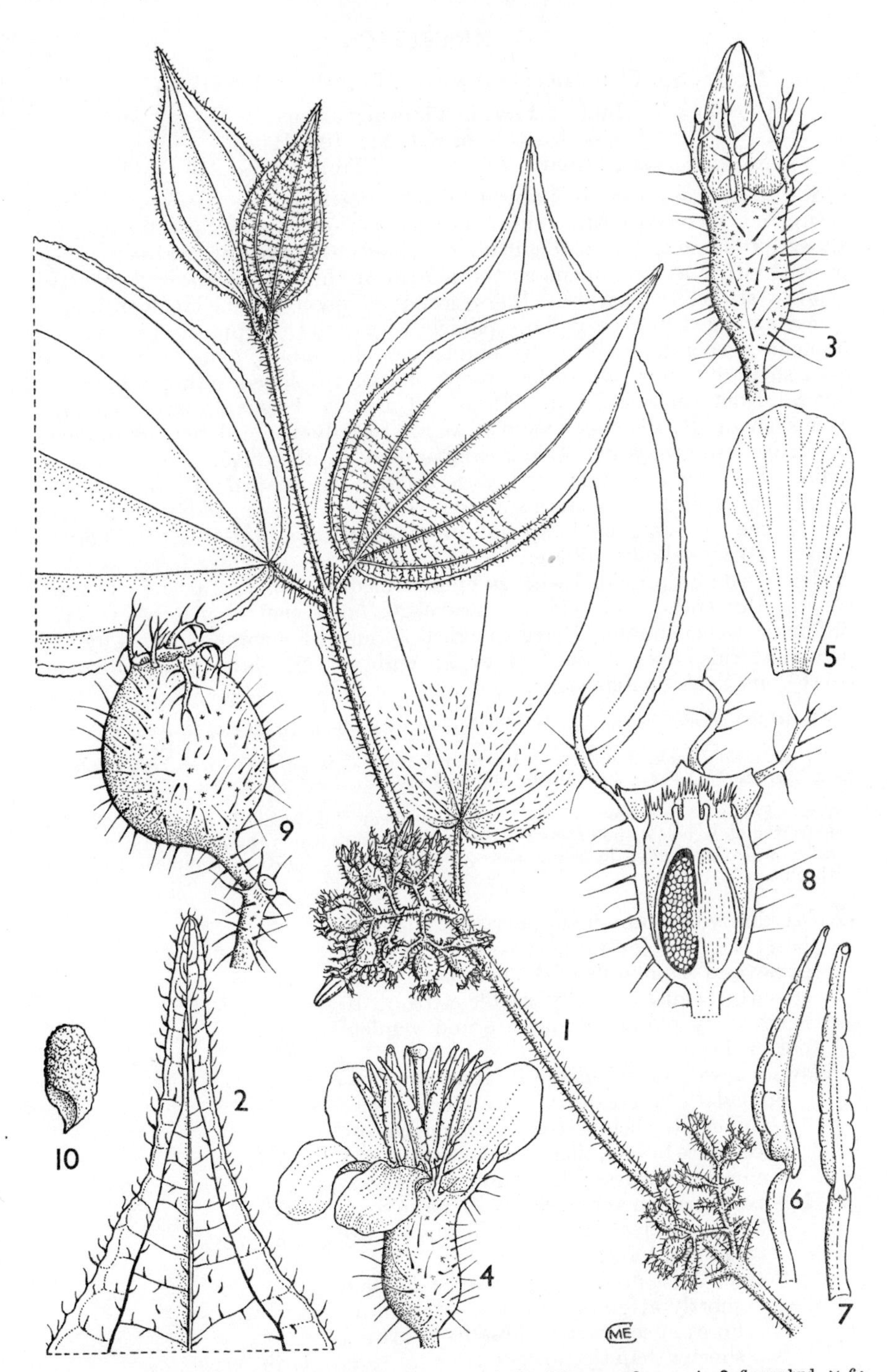

FIG. 20. *CLIDEMIA HIRTA*—**1,** branchlet, × 1; **2,** part of leaf, lower surface, × 4; **3,** flower bud, × 6; **4,** flower, × 4; **5,** petal, × 6; **6, 7,** stamen, side and back views respectively, × 8; **8,** longitudinal section of flower, × 6; **9,** fruit, × 3; **10,** seed, × 24. 1, 3, 8, from *Greenway* 1907; 2, from *Faulkner* 1592; 9, from *McWhirter* 98; 10, from *Faulkner* 4546. Drawn by Mrs. M. E. Church.

17. MEMECYLON

L., Sp. Pl.: 349 (1753) & Gen. Pl., ed. 5: 166 (1754)
Spathandra Guill. & Perr. in Fl. Seneg. Tent. 1: 313 (1833)
Warneckea Gilg in E.J. 34: 100 (1904)
Klaineastrum A. Chev., Veg. Ut. Afr. Trop. Fr. 9: 215 (1917)

Shrubs or trees, usually glabrous, rarely shortly pilose. Leaves opposite, entire, coriaceous to submembranaceous, petiolate or subsessile; midrib always prominent, 2 basal longitudinal nerves either present or absent, when present either as prominent as the midrib or thin and sinuate, transverse veinlets strong to obscure. Inflorescence of pseudo-umbellate contracted cymes arranged in a corymbose panicle with series of opposite or fascicled branches, generally reduced to simpler pseudo-umbels, cymes, fascicles or even single flowers; peduncles absent or well developed, either solitary or several from each leaf-axil. Flowers 4-merous, hermaphrodite (possible exception of *M. dioicum* (Naud.) Cogn.), usually pedicellate; bracts and bracteoles often present. Calyx campanulate to obconical. Petals blue or white. Stamens 8, equal; anthers medifixed, axe-shaped; thecae minute, opening by a slit; connective ending in a thick dorsal appendage as long as the thecae or twice as long and provided with a dorsal sessile hollow gland. Ovary wholly adherent to the calyx-tube, 1-locular, apex swollen or depressed and provided with an epigynous disc with 8 radiating ribs that conceal the anthers in bud; ovules 5–20, on a central placenta; style filiform; stigma simple. Berry spherical, often with a corona formed by the persistent calyx-lobes. Seeds 1 or 2; embryo convolute, exalbuminous; cotyledons thick or foliaceous.

About 320 species in the Old World tropics, with about 70 species in Africa.

M. caeruleum Jack, a native of India and Malaya, is reputed to be cultivated in Zanzibar, see U.O.P.Z.: 350 (1949), but no specimens have been seen.

NOTE. For the purposes of the key, the term "peduncle" refers to all the flowering stalk below the bracts (or ? bracteoles) from which the pedicels (if present) arise and thus may, in some cases, include both the true peduncle and the rhachis.

Leaf-lamina with 3 prominent nerves arising from base, the basal lateral nerves reaching up to or beyond the middle of the lamina; lamina smooth, veinlets usually visible with × 10 lens, rarely papillose-muricate and veinlets not visible:
- Flowers sessile or subsessile in generally epedunculate bracteate inflorescences, rarely pedunculate below the conspicuous inflorescence bracts, then peduncle stout and 5 mm. long or less; leaf-lamina smooth:
  - Inflorescences clustered, each with more than 3 flowers; leaf-lamina cartilaginous, up to 25 cm. long:
    - Petiole 3–10 mm. long; base of lamina shortly attenuate to subcordate; inflorescence-bracts broadly ovate, shorter than the flowers . . . 1. *M. erubescens*
    - Petiole up to 2·5 mm. long; base of lamina truncate to semi-amplexicaul; inflorescence-bracts ovate,

longer than the flowers; short stout peduncle occasionally present below the inflorescence-bracts . . . . 2. *M. amaniense*

Inflorescence solitary, with 1–3 flowers; leaf-lamina membranous, up to 7 cm. long:

Petiole 2–3 mm. long; leaf-lamina ovate, 2–7 cm. long, 1–3·5 cm. wide, base subcordate to rounded . . . . 3. *M. schliebenii*

Petiole 1 mm. long; leaf-lamina broadly ovate, 1·5–1·8 cm. long, 1·2–1·4 cm. wide, base cordate . . . . . 4. *M. microphyllum*

Flowers pedicellate or if subsessile then inflorescence-bracts not conspicuous and peduncle slender, 1–6 mm. long:

Leaf-lamina smooth, not papillose-muricate:

Peduncle present; flowers pedicellate or subsessile:

Flowers 3–5, subsessile; peduncle 1–6 mm. long . . . . . . 5. *M. mouririifolium*

Flowers 1–12; pedicels 4–6 mm. long; peduncle 2–45 mm. long:

Peduncle (including rhachis) 20–45 mm. long, with ± 12 flowers, in the axils of fallen leaves or on old wood . . . . . 7. *M. bequaertii*

Peduncle (including rhachis) 2–7 mm. long, with 1–4-flowers, axillary . 8. *M. jasminoides*

Peduncle absent; flowers pedicellate, in few-many-flowered axillary fascicles . 6. *M. sansibaricum*

Leaf-lamina papillose-muricate:

Inflorescence simple, consisting of an axillary peduncle 2–7 mm. long terminating in 1–3 pedicellate flowers . 10. *M. procteri*

Inflorescence compound, consisting of an axillary peduncle 12–13 mm. long with 0–2 whorls of 1–2 pairs of opposite pedicellate flowers, or occasionally with a pedunculate cyme replacing one of the flowers and terminating in 3 pedicellate flowers . 9. *M. brenanii*

Leaf-lamina with 1 prominent nerve only (the midrib) arising from the base, the lateral basal nerves, if visible, not differing from those arising further up the midrib; lamina always papillose beneath:

Peduncle absent or very short (± 1 mm. long):

Flowers with pedicels 1–5 mm. long; bracts at base of pedicel, not surrounding the calyx:

Leaf-lamina oblong, 24–25 cm. long, 8·5–9·5 cm. wide, base truncate; petiole 2 mm. long . . . . . 11. *M. magnifoliatum*

Leaf-lamina not as above:

Leaf-lamina ovate, ovate-oblong or cordate-oblong, 9–18 cm. long,

3–7 cm. wide, base rounded to subcordate; inflorescences on old wood . . . . . . 12. *M. erythranthum*
Leaf-lamina elliptic, rarely ovate, 5–10·3 cm. long, 2·6–5 cm. wide, base cuneate; inflorescences axillary . . . . 13. *M. semseii*
Flowers subsessile, with pedicels 0·1–0·5 mm. long and 2 bracts immediately below the calyx . . . . 17. *M. myrtilloides*
Peduncle (including rhachis and peduncles where applicable) present, 1 mm. or more long:
Peduncles mostly more than (10–)15 mm. long:
Leaf-lamina more than 3·5 cm. long:
Leaf-lamina ovate to oblong-ovate, 5–10 cm. long, 2–5 cm. wide, long-acuminate, base cordate; petiole ± 1 mm. long; pedicels 3 mm. long . . . . 18. *M. cogniauxii*
Leaf-lamina elliptic, 3·5–12 cm. long, 2·5–6 cm. wide, acuminate, base cuneate; petiole 2–5 mm. long; pedicels 1·5–2 mm. long . . 20. *M. myrianthum*
Leaf-lamina broadly ovate, 1·5–3 cm. long, 1·2 cm. wide, very shortly and obtusely acuminate, base rounded . 19. *M. teitense*
Peduncle less than 10 mm. long:
Pedicels 1 mm. or more long; bracts at the base of the pedicel, not surrounding the calyx:
Leaf-lamina usually more than 3 cm. long, if smaller (2·5 cm.) then longer leaves also present:
Leaf-lamina elliptic, rarely ovate, 5–10·3 cm. long, 2·6–5 cm. wide, broadly acuminate; peduncle 3–4 mm. long; calyx campanulate, 1·2–1·3 mm. long . . 13. *M. semseii*
Leaf-lamina narrowly elliptic to obovate, 2·5–8 cm. long, 1·5–5·5 cm. wide, rounded or emarginate, yellowish green; peduncle 1–10 mm. long; calyx obconic-campanulate, 2 mm. long . . 23. *M. flavovirens*
Leaf-lamina less than 2·7 cm. long:
Leaf-lamina ± as long as wide, broadly ovate or suborbicular, 0·9–2 cm. long, base rounded or broadly cuneate; inflorescence rather dense; peduncle 2–3 mm. long; pedicels 1–2 mm. long . 14. *M. greenwayi*
Leaf-lamina elliptic, 1·3–2·7 cm. long, 0·5–1·6 cm. wide, very shortly and obtusely acuminate, base

cuneate; inflorescence 2–5-flowered; peduncle 2–9 mm. long; pedicels 5–7 mm. long . . . 21. *M. buxoides*

Pedicels 0·1–0·5 mm. long with 2 bracts immediately below the calyx (bracts caducous in *M. fragrans* but scars visible):

Leaves distinctly longer than broad, 0·8–5·5 cm. long; inflorescences either 1(–2)-flowered or else cymes very contracted (with an axis rarely exceeding 3 mm.):

Inflorescences contracted cymes, almost all several-flowered; bracts caducous; leaves ovate-elliptic to lanceolate-elliptic, 2–5·5 cm. long, 1–2·5 cm. wide; fruits ellipsoid . . . . . 15. *M. fragrans*

Inflorescences 1–several in the axils, each 1(–2)-flowered; bracts persistent; leaves smaller or thinner and more ovate:

Fruits globose, 5–6 mm. in diameter; peduncle plus rhachis 5–8 mm. long; leaves ovate-lanceolate to broadly ovate, long-acuminate, 2–5 cm. long, 0·8–3 cm. wide . . . 16. *M. verruculosum*

Fruits ellipsoid, ± 1 cm. long; peduncle plus rhachis 1–3 mm. long; leaves ± rhombic-elliptic, slightly acuminate, 0·8–2 cm. long, 0·6–1 cm. wide . 17. *M. myrtilloides*

Leaves not or little longer than broad, 0·5–1 cm. long; inflorescences cymose, 1–7- or more flowered; fruits globose, 4–5 mm. in diameter . . . . . 22. *M. deminutum*

1. **M. erubescens** *Gilg* in E.M. 2: 41, t. 10/B (1898); V.E. 3(2): 765, fig. 321/A–E (1921); T.T.C.L.: 312 (1949); A. & R. Fernandes in Mem. Soc. Brot. 11: 56 (1956) & in Bol. Soc. Brot., sér. 2, 34: 196 (1960). Type: Tanzania, Lushoto District, Derema [Nderema], *Heinsen* 6 (B, holo.†)

Shrub up to 3(–4·5 *fide* T.T.C.L.) m. high; branchlets obscurely 4-winged when young. Leaf-lamina ovate-lanceolate, ovate or oblong, 6–19 cm. long, 3–9 cm. wide, acuminate, base shortly attenuate to subcordate; basal lateral nerves forming loops with the transverse nerves from about midway to the apex; petiole 3–10 mm. long. Flowers in sessile bracteate axillary fascicles, with tufts of hairs at base of flowers. Calyx-tube campanulate, 2 mm. long, 1·5 mm. in diameter; lobes ovate. Petals obovate-lanceolate, 3·5 mm. long, 2 mm. wide, pink to purple. Anthers 2 mm. long; filaments 4 mm. long. Fruit not known.

TANZANIA. Lushoto District: Amani, 18 Nov. 1947, *Brenan & Greenway* 8338! & 12 Mar. 1950, *Verdcourt* 105! & 25 July 1953, *Drummond & Hemsley* 3454! & 3455!
DISTR. T3; not known elsewhere
HAB. Lowland rain-forest; 870–950 m.

SYN. *M. heinsenii* Gilg in E.M. 2: 42 (1898); V.E. 3(2): 766 (1921); T.T.C.L.: 312 (1949). Type: Tanzania, Lushoto District, Derema, *Heinsen* 6A (B, holo. †)

2. **M. amaniense** (*Gilg*) *A. & R. Fernandes* in Bol. Soc. Brot., sér. 2, 34: 83, t. 21 (1960) & 34: 196 (1960). Type: Tanzania, Lushoto District, Amani, *Warnecke* in *Herb. Amani* 400 (B, holo. †, EA, iso. !)

Evergreen shrub or small tree up to 6 m. high; branchlets obscurely 4-angled or 4-winged when young, later becoming subterete, thickened at the nodes. Leaf-lamina ovate to broadly ovate, 5·5–25 cm. long, 3–16 cm. wide, acuminate, base truncate to semi-amplexicaul, 3–9-nerved from the base, subsessile or petiole up to 2·5 mm. long. Flowers usually in sessile bracteate axillary fascicles, with a tuft of bristles at the base of the flowers, rarely with a short stout peduncle up to 5 mm. long and then flowers conspicuously bracteate. Calyx-tube (immature) cupuliform, ± 1 mm. long, 1·1 mm. in diameter; lobes broadly ovate. Petals oblong-obovate, 3–4 mm. long, 2·5 mm. wide, pink or red. Fruit ellipsoid, 7 mm. in diameter, red.

KENYA. Kwale District: Shimba Hills, *Gardner* in *F.D.* 1409! & Buda Mafisini Forest, 21 Aug. 1953, *Drummond & Hemsley* 3931! & Makadara, 27 Jan. 1959, *Moomaw* 1276!

TANZANIA. Morogoro District: Turiani, 3 June 1933, *B. D. Burtt* 4744! & between Turiani Falls and Mhonda [Mahonda] Sawmill, 4 Nov. 1947, *Brenan & Greenway* 8285! & Morogoro, Aug. 1952, *Semsei* 869.

DISTR. **K7**; **T3**, 6; not known elsewhere

HAB. Lowland rain-forest and riverine forest; 40–600 m.

SYN. *Warneckea amaniensis* Gilg in E.J. 34: 101 (1904); V.E. 3(2): 762, fig. 320 (1921): T.T.C.L.: 313 (1949), excl. specim. *Greenway* 3331; K.T.S.: 265 (1961), incl. specim. *Drummond & Hemsley* 3931

NOTE. More spirit collections of mature flowers and fruits are required in order that this species may be more accurately described and its relationship to *M. erubescens* determined.

3. **M. schliebenii** *Markgraf* in N.B.G.B. 11: 672 (1932); T.T.C.L.: 312 (1949); A. & R. Fernandes in Bol. Soc. Brot., sér. 2, 34: 83, 197 (1960), pro parte, excl. specim. *Drummond & Hemsley* 3808 & *Swynnerton* 32. Type: Tanzania, Ulanga District, Muhulu Mts., *Schlieben* 2083 (B, holo. †, BM, BR, iso. !)

Shrub 2–4 m. high; branchlets angular. Leaf-lamina ovate, 2–7 cm. long, 1–3·5 cm. wide, acuminate, base subcordate to rounded; basal lateral nerves forming loops with transverse nerves from about the middle to the apex; petiole 2–3 mm. long. Inflorescence 1–3-flowered, terminal or axillary. Flowers not seen; data from the original description. Calyx turbinate; lobes obovate, 2 mm. long. Petals broadly elliptic, 2 mm. long, 1·5 mm. wide, pale red. Anthers 1·5 mm. long, filaments 3 mm. long. Fruit subglobose, 7 mm. in diameter.

TANZANIA. Morogoro District: Nguru Mts., near Turiani, 29 Mar. 1953, *Drummond & Hemsley* 1895!; Ulanga District: Muhulu Mts., 14 Apr. 1932, *Schlieben* 2083!

DISTR. **T6**; not known elsewhere

HAB. Upland rain-forest; 1200–1256 m.

NOTE. More material of this species is required, especially from the type locality. The isotypes at the British Museum and Brussels are both sterile, however Markgraf noted that *M. schliebenii* was close to *M. microphyllum* Gilg and *M. erubescens* Gilg (*M. heinsenii*), both with epedunculate, sessile or subsessile inflorescence, from **T3**. *Drummond & Hemsley* 1895 is a fruiting specimen from a locality some 200 miles distant from the type locality yet appears to be of the same species. *Swynnerton* 32, cited by *Fernandes, loc. cit.* (1960), is a sterile specimen from Sokoke in **K7** and is probably *M. mouririifolium* Brenan.

Magogo & Glover 1047 from Lango ya Mwagandi [Longo Mwagandi] Forest, Shimba Hills, Kwale District, may belong here. It differs in having longer acuminate leaves up to 8 cm. long, 4·5 cm. wide. It may be that when more material becomes available this species will be found to be a variant of *M. amaniense*.

4. **M. microphyllum** *Gilg* in E.J. 34: 101 (1904); Engl. in V.E. 1(1): 312, fig. 279/F–M (1910) & 3(2): 766, fig. 321/F–M (1921), descr. ampl.; T.T.C.L.: 312 (1949); A. & R. Fernandes in Bol. Soc. Brot., sér. 2, 34: 198 (1960). Type: Tanzania, Lushoto District, Amani, *Warnecke* in *Herb. Amani* 459 (B, holo. †)

Shrub 2–3 m. high. Leaf-lamina broadly-ovate, 1·5–1·8 cm. long, 1·2–1·4 cm. wide, acuminate, base cordate; petiole ± 1 mm. long. Inflorescence of 1–3-flowered subsessile axillary clusters; pedicel 1 mm. long. Calyx cupuliform, ± 1 mm. long, 1·1 mm. in diameter; lobes ovate. Petals oblong-lanceolate, 3·7 mm. long, 1 mm. wide, pink. Anthers 1·4 mm. long; filaments 2·4 mm. long. Fruit not known.

TANZANIA. Lushoto District: Kwamkoro–Sangerawe, 23 Mar. 1917, *Zimmermann* in *Herb. Amani* 6703!
DISTR. **T3**; not known elsewhere
HAB. Lowland rain-forest; ± 900 m.

NOTE. This species is close to *M. sansibaricum* Taub. and to *M. mouririifolium* Brenan, from which it may readily be distinguished by the characters given in the key. Further gatherings of *M. microphyllum* are required.

5. **M. mouririifolium** *Brenan* in K.B. 1: 91, fig. 1/1 (1947); A. & R. Fernandes in Mem. Soc. Brot. 11: 57 (1956) & in Bol. Soc. Brot., sér. 2, 34: 197 (1960); K.T.S.: 264 (1961). Type: Kenya, Kilifi District, Arabuko, *R. M. Graham* in *F.D.* 1991 (K, holo.!, EA, FHO, PRE, iso.)

Evergreen shrub up to 4·5 m. high; branchlets obscurely 4-angled. Leaf-lamina ovate to broadly ovate, 1·2–5(–8) cm. long, 0·6–2·7(–4·5) cm. wide, subacuminate, base rounded to subcordate; 2 basal lateral nerves scarcely raised beneath; petiole 1–2·5 mm. long. Inflorescence pedunculate axillary clusters of 3–5 subsessile flowers; peduncle 1–6 mm. long. Calyx-tube broadly obconical, 2 mm. long, 2·2–2·5 mm. in diameter; calyx-lobes semi-orbicular. Petals obovate, 1·7–2·5 mm. long, 1·2–1·6 mm. wide, white or cream. Anthers 1–1·2 mm. long; filaments 2–3·8 mm. long. Fruit ovoid, 7 mm. long, 6 mm. in diameter.

KENYA. Kwale District: Lungalunga–Msambweni road between Umba R. and Mwena R., 18 Aug. 1953, *Drummond & Hemsley* 3857!; Kilifi District: Arabuko Forest, *R. M. Graham* in *F.D.* 1527! & Jilori, 25 Nov. 1961, *Polhill & Paulo* 852!
DISTR. **K7**; ?**T3**; not known elsewhere
HAB. Lowland dry evergreen forest and rain-forest; 45–100 m.

SYN. [*M. schliebenii* sensu A. & R. Fernandes in Bol. Soc. Brot. sér. 2, 34: 83, 197 (1960), pro specim. *Drummond & Hemsley* 3808, *non* Markgraf]

NOTE. Two specimens from wetter forests in the Lushoto District, *Brenan* 8357 from Sangerawe and *Harris* 103A from Dindira, may belong here but lack flowers; the presence of a 6 mm. long "peduncle" distinguishes the specimens from *M. microphyllum* Gilg which has been recorded from the same locality.

6. **M. sansibaricum** *Taub.* in P.O.A. C: 296 (1895); Gilg in E.M. 2: 40 (1898); V.E. 3(2): 765 (1921); A. & R. Fernandes in Bol. Soc. Brot., sér. 2, 34: 195 (1960) & 43: 302 (1969). Type: Zanzibar I., Jambiani [Jambiana], *Stuhlmann*, ser. 1, 718 (B, holo. †)

Evergreen shrub or small tree up to 15 m. high; branchlets obscurely 4-angled. Leaf-lamina ovate to elliptic, sometimes broadly so, 1·4–

9·5(–12) cm. long, 1·2–5(–6·4) cm. wide, apex obtuse to long-acuminate, base cuneate to obtuse; midrib and 2 basal lateral nerves subprominent beneath; vein-reticulation conspicuous (at least when dried); petiole 1–4 mm. long. Inflorescences subsessile axillary 1–many-flowered fascicles; pedicels 3–17 mm. long. Calyx obconical to obconical-campanulate, 1–1·5 mm. long, 1·3–1·5(–3) mm. in diameter; lobes short. Petals oblong-lanceolate, rhombic or obovate, 1·5–3 mm. long, 0·5–2·8 mm. wide, white, turning blue. Anthers 0·5–1·5 mm. long; filaments 3–6 mm. long. Fruit 5–8 mm. in diameter, cobalt blue.

var. **sansibaricum**; *A. & R. Fernandes* in Bol. Soc. Brot., sér. 2, 43: 302 (1969), pro parte, excl. syn. *M. buchananii* Gilg

Shrub or small tree up to 7(–10·5) m. high. Leaf-lamina broadly ovate, obtuse to shortly acuminate.

KENYA. Kilifi District: Arabuko, Apr. 1930, *Donald* 1 in *F.D.* 2327! & Sokoke Forest, 14 Apr. 1954, *Trump* 136! & *Gisau* in *E.A.H.* 10988!
TANZANIA. Morogoro District: Lusunguru Forest Reserve, 19 Oct. 1959, *Mgaza* 333!; Kilwa District: Selous Game Reserve, Libungani, 22 Feb. 1971, *Ludanga* 1258!; Lindi District: Lutamba Lake, 4 Dec. 1934, *Schlieben* 5688!; Zanzibar I., Panga Juu Cave-well, 10 June 1931, *Vaughan* 1339!
DISTR. **K7**; **T3**, 6, 8; **Z**; Mozambique
HAB. Lowland dry evergreen forest, deciduous woodland; 10–100 m.

SYN. *M. lutambense* Markgraf in N.B.G.B. 12: 717 (1935); T.T.C.L.: 312 (1949). Type: Tanzania, Lindi District, Lutamba Lake, *Schlieben* 5688 (B, holo. †, K, iso.!)
M. melindense A. & R. Fernandes in Bol. Soc. Brot., sér. 2, 29: 62, t. 16 (1955) & in Mem. Soc. Brot. 11: 57, t. 7 (1956) & in Bol. Soc. Brot., sér. 2, 34: 77, 195 (1960); K.T.S.: 264 (1961). Type: Kenya, Kilifi District, Arabuko, *Dale* in *F.D.* 3835 (PRE, holo., EA, K, iso.!)
M. buchananii Gilg var. *maritimum* A. & R. Fernandes in Bol. Soc. Brot., sér. 2, 34: 77, t. 16 (1960) & 34: 195 (1960). Type: Kenya, Lamu District, Utwani, *Mohamed Abdullah* in *F.D.* 3351 (EA, holo., K, iso.!)
M. sp. nov. sensu K.T.S.: 264 (1961), pro specim. *Donald* 2327 & *Mohamed Abdullah* in *F.D.* 3351
M. sansibaricum Taub. var. *maritimum* (A. & R. Fernandes) A. & R. Fernandes in Bol. Soc. Brot., sér. 2, 43: 304 (1969)
M. sousae A. & R. Fernandes in Bol. Soc. Brot., sér. 2, 46: 67, t. 4 (1972). Type: Mozambique, Beira, *Gomes & Sousa* 4380 (COI, holo., K, iso.!)

var. **buchananii** (*Gilg*) *A. & R. Fernandes* in Bol. Soc. Brot., sér. 2, 46: 66 (1972). Type: Malawi, *Buchanan* 141 (B, holo. †, K, iso.!)

Shrub or small tree up to 10 m. high. Leaf-lamina oblong-ovate, long acuminate.

TANZANIA. Rungwe District: Kiwira R., 7 Oct. 1912, *Stolz* 1583! & 28 Nov. 1912, *Stolz* 1705!
DISTR. **T7**; Mozambique, Malawi and Zambia
HAB. Lowland rain-forest, woodland; 900 m.

SYN. *M. buchananii* Gilg in E.M. 2: 40 (1898); A. & R. Fernandes in Bol. Soc. Brot. sér. 2, 34: 76 (1960), pro parte; F.F.N.R.: 310 (1962)
M. stolzii Engl., V.E. 3(2): 765 (1921); T.T.C.L.: 312 (1949); A. & R. Fernandes in Mem. Soc. Brot. 11: 55 (1956). Type: Tanzania, Rungwe District, Kiwira R. [Kibila R.], *Stolz* 1583 (B, holo. †, BM!, EA, K!, PRE, iso.)
[*M. sansibaricum* Taub. var. *sansibaricum* sensu A. & R. Fernandes in Bol. Soc. Brot., sér. 2, 43: 309 (1969), pro parte]

NOTE. A rather variable species in which a coastal and an inland variety are recognized here. A small leaved form known as *M. melindense* A. & R. Fernandes is here regarded as a shrub or possibly juvenile form of var. *sansibaricum*; both the small and typical leaf form are known from the same localities. Further gatherings and field studies may prove the two forms to be separable.

Similarly *M. sansibaricum* var. *maritimum* (A. & R. Fernandes) A. & R. Fernandes is regarded as being within the acceptable range of variation for var. *sansibaricum*; again further collections and field studies are required.

7. **M. bequaertii** *De Wild.* in Rev. Zool. Afr. 9, Suppl. Bot.: 14 (1921) & Pl. Bequaert. 1: 388 (1922); A. & R. Fernandes in Bol. Soc. Brot., sér. 2, 34: 78, 196 (1960). Type: Zaire, Nandefu, between Penghe and Irumu, *Bequaert* 2666 (BR, holo.)

Tree 9–15 m. high; branchlets obscurely 4-winged. Leaf-lamina elliptic, 9–15·5(–19) cm. long, 3–7(–9) cm. wide; 8–9 pairs of transverse lateral nerves ± prominent beneath and forming loops with the 2 basal lateral nerves from about midway to the apex; petiole 2–3 mm. long. Inflorescence ± 12-flowered, axillary, on the older wood; peduncle (with rhachis) ± 20(–45) mm. long, forked; pedicels 4–5 mm. long. Calyx obconical, ± 2 mm. long, 1·2 mm. in diameter; lobes triangular-ovate, ± 1 mm. long. Petals obovate, 1·5–2 mm. long, 2 mm. wide, bluish-pink. Anthers 1 mm. long; filaments not developed. Fruit globose, 9–10 mm. in diameter.

UGANDA. Kigezi District: Kigezi, *Butt* K24! & Impenetrable Forest, Apr. 1948, *Purseglove* 2679!
DISTR. **U**2; Zaire
HAB. Rain-forest, sometimes the dominant understory tree; 1500 m.

NOTE. Further collections of the species are required as well as information on whether the plant is truly cauliflorous.

8. **M. jasminoides** *Gilg* in E.M. 2: 39 (1898); V.E. 3(2): 764 (1921); T.T.C.L.: 312 (1949); I.T.U., ed. 2: 170 (1952), excl. syn. *M. strychnoides* Gilg; A. & R. Fernandes in Mem. Soc. Brot. 11: 54 (1956) & in Bol. Soc. Brot., sér. 2, 34: 74, t. 14 (1960) & 34: 193 (1960). Type: Zaire, Bongwa's village, Apr. 1870, *Schweinfurth* 3609 (B, holo. †, K, iso.!)

Evergreen shrub or tree up to 15 m. high; branchlets acutely 4-angled. Leaf-lamina ovate to elliptic, 4–15 cm. long, 2–6·5 cm. wide; transverse nerves scarcely prominent beneath and not forming conspicuous loops with the basal lateral nerves; petiole 2–5 mm. long. Inflorescence 1–3(–4)-flowered, axillary; peduncles 2–7 mm. long; pedicels 4–6 mm. long. Calyx obconical, ± 1·5 mm. long; lobes broadly triangular, 1 mm. long. Petals semicircular, 2 mm. long, 3 mm. wide, bright blue. Anthers 1 mm. long; filaments 3 mm. long. Fruit globose, ± 7 mm. in diameter, changing from green to pale blue to very dark blue to almost black at maturity.

UGANDA. Bunyoro District: Budongo Forest, Apr. 1932, *Harris* 99!; Masaka District: Sese Is., Sozi, Dec. 1922, *Maitland* 385!; Mengo District: Mpanga Forest Reserve, 11 Oct. 1953, *Drummond & Hemsley* 4741!
TANZANIA. Bukoba District: Minziro Forest, July 1951, *Eggeling* 6250! & Kantale, *Gillman* 299!; Kigoma District: Kasakati, 13 Sept. 1965, *Suzuki* B-34!
DISTR. **U**2, 4; **T**1, 4; Zaire
HAB. Lowland rain-forest and swamp forest; 1080–1170 m.

SYN. [*M. membranifolium* sensu Taub. in P.O.A. C: 296 (1895), *non* Hook. f.]
M. heterophyllum Gilg in E.M. 2: 39 (1898); V.E. 3(2): 764 (1921); T.T.C.L.: 312 (1949). Types: Tanzania, Bukoba, *Stuhlmann* 957 & 1549 (both B, syn. †)
M. wilwerthii De Wild. in Ann. Mus. Congo., Bot., sér. 5, 3: 246, t. 32/1 (1910). Type: Zaire, Uputo, 1896, *Wilwerth* (BR, holo.)
M. cyaneum De Wild. in Rev. Zool. Afr. 9, Suppl. Bot. 1: 15 (1921) & Pl. Bequaert. 1: 389 (1922). Types: Zaire, Avakubi, *Bequaert* 2003 & Irumu, *Bequaert* 2913 & Semliki, *Bequaert* 3171 (all BR, syn.)

NOTE. *Synnott* 569 from Budongo Forest appears to be an epedunculate form of this species.

9. **M. brenanii** *A. & R. Fernandes* in Bol. Soc. Brot., sér. 2, 34: 71, t. 10 (1960) & 34: 193 (1960). Type: Tanzania, Lushoto District, Sangerawe, 1934, *Greenway* 3680 (EA, holo., K, iso.!)

Evergreen tree up to 18 m. high; branchlets 4-angled. Leaf-lamina broadly elliptic, 4–7·5 cm. long, 2·3–4 cm. wide, obtusely acuminate, base cuneate, both surfaces densely papillose-muricate; petiole 2–4 mm. long. Inflorescence a 3–11-flowered axillary cyme; peduncle (with rhachides) 12–15 mm. long; pedicels 2–4 mm. long. Calyx broadly obconical, 1·5 mm. long, 2·5 mm. in diameter; lobes semicircular, 1 mm. long. Petals oblong to obovate, 3 mm. long, 2 mm. wide, white. Anthers 1·5 mm. long; filaments 4 mm. long. Fruit (immature) globose, 7 mm. in diameter.

TANZANIA. Lushoto District: Sangerawe, 4 Jan. 1934, *Greenway* 3680! & Kwamkoro Forest Reserve, 11 Aug. 1961, *Mgaza* 437!
DISTR. **T3**; not known elsewhere
HAB. Lowland rain-forest; 1050 m.

SYN. *M. sp.* sensu T.T.C.L.: 313 (1949)

10. **M. procteri** *A. & R. Fernandes* in Bol. Soc. Brot., sér. 2, 34: 72, t. 11 (1960) & 34: 193 (1960). Type: Tanzania, Lushoto District, Shagayu, *Procter* 208 (EA, holo., K, iso.!)

Evergreen shrub up to 3 m. high; branchlets obscurely 4-winged. Leaf-lamina broadly elliptic, 2·5–5 cm. long, 1·5–3·2 cm. wide, apex obtuse or occasionally rounded, base cuneate, both surfaces densely papillose-muricate; petiole 3–5 mm. long. Inflorescence a 1–3-flowered axillary cyme; peduncle single, 2–7 mm. long; pedicels 2–4 mm. long. Flowers not seen. Fruit (immature) globose, 8 mm. in diameter.

TANZANIA. Lushoto District: Shagayu, May 1953, *Procter* 208! & 1 Oct. 1964, *Mgaza* 616!
DISTR. **T3**; not known elsewhere
HAB. Upland rain-forest; 1950 m.

NOTE. Flowering material of this species is required. Perhaps this is only a shrub form of *M. brenanii* A. & R. Fernandes, but there is insufficient material of both species for any taxonomic decision to be made.

11. **M. magnifoliatum** *A. & R. Fernandes* in Bol. Soc. Brot., sér. 2, 43: 305, t. 17 (1969). Type: Tanzania, Ulanga District, Magombera Forest Reserve, *Semsei* 3375 (COI, holo.!, EA, K, iso.!)

Habit not recorded. Leaf-lamina oblong, 24–25 cm. long, 8·5–9·5 cm. wide, shortly acuminate, base truncate, with 25–28 pairs of inconspicuous transverse nerves, minutely papillose on both surfaces; petiole 2 mm. long. Inflorescence ? few-flowered, cauliflorous; peduncle absent; pedicels 5 mm. long. Flowers not known. Fruit obovoid, 17–20 mm. long, 12–14 mm. in diameter, blue-black.

TANZANIA. Ulanga District: Magombera Forest Reserve, 2 Nov. 1961, *Semsei* 3375!
DISTR. **T6**; not known elsewhere
HAB. Lowland rain-forest; 200–250 m.

NOTE. More material is required of this unusual species, which might prove to be no more than a luxuriant form of *M. erythranthum*

12. **M. erythranthum** *Gilg* in E.M. 2: 45, t. 10/C (1898); V.E. 3(2): 769 (1921); T.T.C.L.: 311 (1949); A. & R. Fernandes in Mem. Soc. Brot. 11: 60 (1956) & in Bol. Soc. Brot., sér. 2, 34: 201 (1960), excl. specim. *Greenway* 4814 & *Zimmermann* 6704 & 6705. Type: Tanzania, Lushoto District, Derema [Nderema], *Heinsen* 3 (B, holo. †, K, iso.!)

Evergreen shrub or tree up to 8 m. high. Leaf-lamina ovate, ovate-oblong or cordate-oblong, 9–18 cm. long, 3–7 cm. wide, obtusely acuminate, base rounded to subcordate, with the midrib impressed above, prominent beneath, and 18–24 transverse nerves faintly discernible on both surfaces, minutely papillose on both surfaces, glossy above; petiole 2–4 mm. long. Inflorescences sessile, ± 20–40-flowered clusters on old wood; pedicels 1·5–3 mm. long. Calyx obconical, ± 1·5 mm. long, 1·3 mm. in diameter; lobes deltoid ± 0·5 mm. long. Petals subcircular, ± 1 mm. long, ± 1 mm. wide, pink to scarlet. Stamens not seen. Fruit subglobose, ± 10 mm. in diameter, red becoming black when mature.

TANZANIA. Lushoto District: Derema, *Heinsen* 3!; Tanga District: Mlinga Peak, 18 Feb. 1937 & 2 Feb. 1939, *Greenway* 4904! & 5848!; Morogoro District; Uluguru Mts., 17 May 1938, *Schlieben* 3950!
DISTR. **T3**, 6; Rhodesia
HAB. Rain-forest; 750–1300 m.

13. **M. semseii** *A. & R. Fernandes* in Bol. Soc. Brot., sér. 2, 43: 300, t. 14 (1969), as "*semsei*". Type: Tanzania, Lushoto District, Kwamkoro, *Semsei* 2956 (EA, holo., K, iso.!)

Evergreen shrub or small tree up to 4·5 m. high (or ? more). Leaf-lamina elliptic, rarely ovate, 5–10·3 cm. long, 2·6–5 cm. wide, apex broadly acuminate, rarely retuse, base cuneate, yellowish green; transverse nerves 10–16, faint; petiole 2–4 mm. long. Inflorescence a sessile or rarely shortly pedunculate many-flowered dense axillary fascicle; peduncle 3–4 mm.; pedicels 1–3·5 mm. long, in fruit ± 5 mm. Calyx campanulate, 1·2–1·3 mm. long, 1 mm. in diameter; lobes deltoid, ± 0·5 mm. long. Petals triangular, ± 2 mm. long, 1·5 mm. wide, pale rose. Anthers ± 1·5 mm. long; filaments 3·3 mm. long. Fruit (immature) globose, ± 7 mm. in diameter, black.

TANZANIA. Lushoto District: Maramba, 23 Jan. 1917, *Zimmermann* in *Herb. Amani* 6705! & Kwamkoro–Potwe, 31 Dec. 1936, *Greenway* 4814! & 2 Mar. 1960, *Titila Msuya* 18!
DISTR. **T3**; not known elsewhere
HAB. Lowland rain-forest; 900–960 m.

SYN. [*M. erythranthum* sensu A. & R. Fernandes in Bol. Soc. Brot., sér. 2, 34: 201 (1960), quoad specim. *Greenway* 4814 & *Zimmermann* 6704 & 6705, *non* Gilg]

NOTE. Related to *M. erythranthum* from which it is distinguished by the characters given in the key.

14. **M. greenwayi** *Brenan* in K.B. 1: 95, fig. 2/2 (1947); T.T.C.L.: 311 (1949); A. & R. Fernandes in Mem. Soc. Brot. 11: 59 (1956) & in Bol. Soc. Brot., sér. 2, 34: 200 (1960). Type: Tanzania, Tanga District, Mlinga Peak, *Greenway* 6632 (K, holo.!, EA, FHO, iso.)

Evergreen tree up to 12 m. high. Leaf-lamina broadly ovate or suborbicular, 0·9–2 cm. long, 0·9–1·9 cm. wide, apex broadly obtuse or rounded and often emarginate, base rounded or broadly cuneate; petiole 1·5–2·5 mm. long. Inflorescence axillary, many flowered, consisting of clusters of solitary flowers and pedunculate pseudo-umbellate cymes; peduncle 2–3 mm. long or rudimentary; pedicels 1–2 mm. long. Calyx-tube semi-globose below, constricted in the middle, campanulate above, 1·7–2 mm. long, 2–2·3 mm. in diameter; lobes broadly triangular, ± 0·6 mm. high. Petals broadly ovate, 1·8 mm. long, 2 mm. wide, pale pink to almost white. Anthers ± 1·3 mm. long; filaments 3·8 mm. long. Fruit not known.

TANZANIA. Tanga District: Mlinga Peak, 24 July 1932, 2 Feb. 1939 & 28 Aug. 1942, *Greenway* 2999!, 5851! & 6632!
DISTR. **T3**; known only from Mt. Mlinga
HAB. Locally dominant in rain-forest; 1050–1080 m.

15. **M. fragrans** *A. & R. Fernandes* in Bol. Soc. Brot., sér. 2, 34: 87, t. 26 (1960) & 34: 199 (1960). Type: Kenya, Kilifi District, Sokoke Forest, *Jeffery* 616 (EA, holo. !)

Usually a shrub up to 2·5 m. high, rarely a tree to 12 m. Leaf-lamina ovate-elliptic to lanceolate-elliptic, 2–5·5 cm. long, 1–2·4 cm. wide, acuminate to an obtuse apex, base rounded to broadly cuneate, chartaceous to thinly coriaceous, muricate-papillate beneath. Inflorescences terminal and axillary, shortly cymose, almost always several-flowered; peduncle 1–2 mm. long; rhachides 0·5–1·5 mm. long; pedicels ± 0·5 mm. long; bracts caducous. Calyx obconic, ± 2·5 mm. long and broad; lobes very broadly triangular, ± 0·5 mm. long. Petals subtriangular, 2·5 mm. long, 1·5 mm. wide, white. Anthers 1·8 mm. long; filaments 4 mm. long. Fruit ellipsoid, 8–9 mm. long, 6–7 mm. in diameter, black.

KENYA. Kwale District: Lungalunga–Msambweni, between Umba and Mwena Rivers, 18 Aug. 1953, *Drummond & Hemsley* 3859!; Kilifi District: Kikuyuni, 26 Dec. 1954, *Verdcourt* 1180! & S. of Jilore Forest Station, 20 Nov. 1969, *Perdue & Kibuwa* 10010!

DISTR. **K7**; not known elsewhere

HAB. Coastal dry evergreen forest; 60–100 m.

SYN. [*M. verruculosum* sensu K.T.S.: 264 (1961), pro specim. *Drummond & Hemsley* 3859, *non* Brenan]

NOTE. Easily distinguished from *M. verruculosum* by the short several-flowered inflorescences, ellipsoid fruits and thicker leaves.

16. **M. verruculosum** *Brenan* in K.B. 1: 92, fig. 2/1 (1947); A. & R. Fernandes in Mem. Soc. Brot. 11: 59 (1956) & in Bol. Soc. Brot., sér. 2, 34: 87, 199 (1960); K.T.S.: 264 (1961), excl. specim. *Drummond & Hemsley* 3859. Type: Kenya, Kwale District, Makadara, *R. M. Graham* C. 571 in *F.D.* 2040 (K, holo. !, BM !, EA, FHO, PRE, iso.)

Shrub up to 4·5 m. high. Leaf-lamina ovate-lanceolate to ovate to broadly ovate, 2–5 cm. long, 0·8–3 cm. wide, long-acuminate, apex acute to obtuse, base broadly cuneate to rounded, thinly chartaceous, papillose-muricate on both surfaces; petiole 0–1·5 mm. long. Inflorescences 1-several from the axils, 1(–2)-flowered; peduncle 1–4 mm. long; rhachis 3–6 mm. long; pedicel 0·1–0·5 mm. long; bracts linear-lanceolate. Calyx-tube obconical, 1·5–2 mm. long and broad; lobes broadly triangular, 0·5 mm. long. Petals elliptic, 2·5–3·5 mm. long, 1·5–2 mm. wide, white. Anthers 1·3 mm. long; filaments 4 mm. long. Fruit globose, 5–6 mm. in diameter, greenish yellow, becoming blackish red.

KENYA. Kwale District: Shimba Hills, *Gardner* in *F.D.* 1447! & Mwele Mdogo Forest, 23 Aug. 1953, *Drummond & Hemsley* 3971! & Marere Hill, 7 Mar. 1968, *Magogo & Glover* 237!

TANZANIA. Morogoro District: Turiani Falls, 8 Dec. 1935, *B. D. Burtt* 5418! & Turiani, Nov. 1953, *Semsei* 1477! & Mtibwa Forest Reserve, Aug. 1952, *Semsei* 901!

DISTR. **K7**; **T6**; not known elsewhere

HAB. Riverine forest and lowland evergreen forest; 300–750 m.

NOTE. Closely related to *M. myrtilloides* Markgraf, but differing in the smaller globose fruits, which appear to ripen blackish red, not blue, by the longer inflorescence-axes and the larger thinner more ovate leaves. Specimens from Tanzania have consistently shorter inflorescence-axes than those from Kenya, but the material available is insufficient to show whether this difference warrants formal taxonomic recognition.

17. **M. myrtilloides** *Markgraf* in N.B.G.B. 11: 1078 (1934); T.T.C.L.: 312 (1949); A. & R. Fernandes in Bol. Soc. Brot., sér. 2, 34: 199 (1960). Type: Tanzania, Morogoro District, Mkambaku Mt., *Schlieben* 3581 (B, holo. †, BM, iso. !)

Shrub or small tree 1·5–3 m. tall. Leaf-lamina ± rhombic-elliptic, 0·8–2 cm. long, 0·6–1 cm. wide, slightly acuminate, apex acute or obtuse, base rather long-cuneate, papillate-muricate on both surfaces; petiole 0·5–1 mm. long. Inflorescences 1-flowered, axillary; peduncle 1–1·5 mm. long; rhachis 0·5–1 mm. long; pedicel negligible. Calyx obconical, 1–1·5 mm. long, 1–2·5 mm. in diameter; lobes shallowly deltoid. Petals 1–2 mm. long, white. Anthers and filaments 1·5 mm. long. Fruit ellipsoid, ± 1 cm. long and 0·6 cm. in diameter, ultramarine blue.

TANZANIA. Morogoro District: Bunduki, Salaza Forest, 15 Mar. 1953, *Drummond & Hemsley* 1610! & Lupanga, 21 Nov. 1969, *Pócs & Csontos* 6066/c! & Uluguru Mts., without precise locality, 27 Dec. 1938, *Vaughan* 2675!

DISTR. **T6**; known only from the Uluguru Mts.

HAB. Upland rain-forest; 1350–2000 m.

18. **M. cogniauxii** *Gilg* in E.M. 2: 44, t. 10/A (1898); V.E. 3 (2): 768 (1921); T.T.C.L.: 311 (1949); A. & R. Fernandes in Mem. Soc. Brot. 11: 60 (1956) & in Bol. Soc. Brot., sér. 2, 34: 200 (1960). Type: Tanzania, Lushoto District, Derema [Nderema], *Heinsen* 5 (B, holo. †, BM!, EA, iso.)

Evergreen shrub or small tree up to 6 m. high. Leaf-lamina ovate to oblong-ovate, 5–10 cm. long, 2–5 cm. wide, long-acuminate, base cordate; petiole ± 1 mm. long. Inflorescence a (1–)5–12-flowered umbellate cluster on a solitary axillary peduncle 2–3 cm. long; pedicels 3(–5 in fruit) mm. long. Calyx-tube obconical, 1·5 mm. long, 1·5 mm. in diameter; lobes shallowly deltoid, 0·5 mm. high. Petals obovate, 2 mm. long, 1·5 mm. wide, cream or pink. Anthers 1·3 mm. long; filaments 2·5 mm. long. Fruit subglobose, 7–8 mm. in diameter, blue.

TANZANIA. Lushoto District: Amani, 24 Feb. 1950, *Verdcourt* 86! & Kwamkoro, 14 Dec. 1959, *Semsei* 2954!; Morogoro District: Tegetero, 20 Mar. 1953, *Drummond & Hemsley* 1721!

DISTR. **T3, 6**; not known elsewhere

HAB. Rain-forest; 750–1800 m.

19. **M. teitense** *Wickens* in K.B. 29: 148, fig. 5 (1974). Type: Kenya, Teita District, Mbololo Hill, *Faden et al.* 71/45 (K, holo.!, EA, iso.!)

Shrub or tree to 22 m. high. Leaf-lamina broadly ovate, 1·5–3 cm. long, 1–2 cm. wide, very shortly and obtusely acuminate, base rounded; petiole ± 0·5 mm. long. Inflorescence 6–12(or ? more)-flowered pseudo-umbellate contracted cymes; peduncle (7–)10–15 mm. long; pedicels 2–3 mm. long. Calyx-tube shallowly obconical, 1–1·2 mm. high, 1·5–2·2 mm. in diameter; lobes shallowly deltoid, 0·3 mm. high. Petals broadly ovate, 1·5 mm. long, 1·5 mm. wide, white. Anthers ± 1 mm. long; filaments 1·5–2 mm. long. Fruit not seen. Fig. 21.

KENYA. Teita District: Ngangao, 15 Sept. 1953, *Drummond & Hemsley* 4363! & Mbololo Hill, Mraru Ridge, 2 Jan. 1971, *Faden et al.* 71/45!

DISTR. **K7**; not known elsewhere

HAB. Upland rain-forest; 1440–1950 m.

NOTE. Related to *M. greenwayi* Brenan from which it is readily identified by the "pedunculate" umbellate inflorescence.

20. **M. myrianthum** *Gilg* in E.M. 2: 44 (1898); V.E. 3 (2): 768 (1921); T.T.C.L.: 312 (1949); I.T.U., ed. 2: 170 (1952); A. & R. Fernandes in Bol. Soc. Brot., sér. 2, 34: 84, 198 (1960) & in C.F.A. 4: 118 (1970); F.F.N.R.: 435 (1962). Type: Angola, Pungo Andongo, *Welwitsch* 911 (LISU, lecto., BM!, COI, K!, isolecto.)

Evergreen tree up to 9 m. high; branchlets bluntly 4-angled. Leaf-

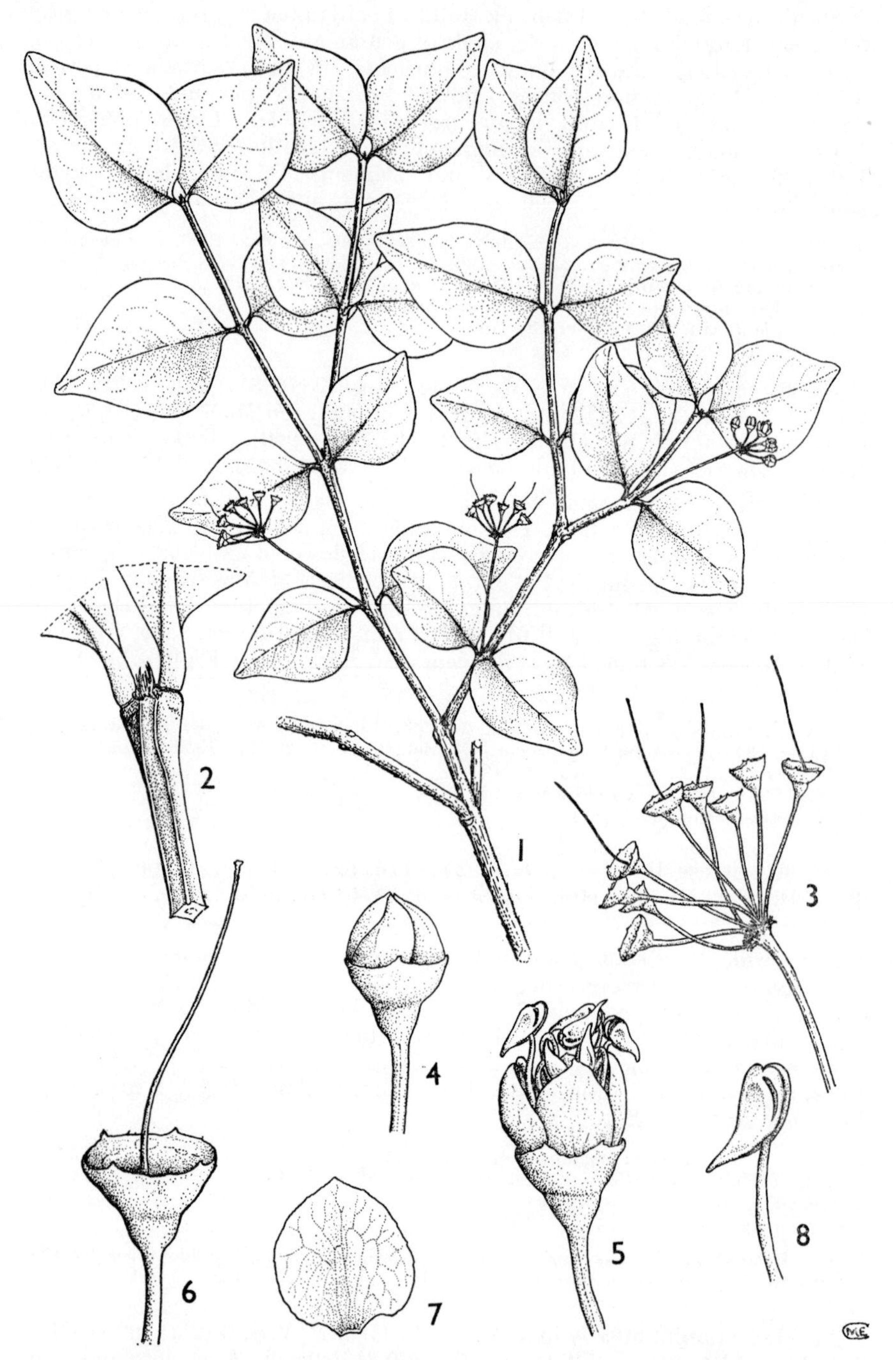

FIG. 21. *MEMECYLON TEITENSE*—**1,** flowering branch, × 1; **2,** apex of young shoot, × 4; **3,** inflorescence, × 6; **4,** flower bud, × 8; **5,** flower, × 6; **6,** flower after fall of petals and stamens, × 10; **7,** petal, × 10; **8,** stamen, × 12. All from *Faden* 71/45. Drawn by Mrs. M. E. Church.

lamina elliptic, 3·5–8(–12) cm. long, 2·5–5(–6) cm. wide, acuminate, base cuneate; petiole 2–5 mm. long. Inflorescences many-flowered compound pedunculate pseudo-umbels; rhachis 15–27 mm. long, peduncles 5–14 mm. long; pedicels 1·5–2 mm. long. Calyx-tube obconical, ± 1·3 mm. long, 1·5 mm. in diameter; calyx shallowly sinuate. Petals broadly ovate, 1·5 mm. long, 1·7 mm. wide, white. Anthers ± 1 mm. long, vivid blue; filaments ± 1·5 mm. long. Fruit globose, ± 8 mm. in diameter.

UGANDA. Masaka District: Minziro Forest, July 1938, *Eggeling* 3755! & Bugala I., June 1951, *Philips* 477! & Malabigambo Forest, 2 Oct. 1953, *Drummond & Hemsley* 4536!
TANZANIA. Bukoba District: Munene, Sept./Oct. 1935, *Gillman* 613!
DISTR. **U4**; **T1**; Gabon, Zaire, Zambia and Angola
HAB. Lowland rain-forest; 1100–1200 m.

SYN. *M. claessensii* De Wild. in B.J.B.B. 4: 426 (1914). Type: Zaire, Katako-Komba, 1910, *Claessens* 402 (BR, holo.)

21. **M. buxoides** *Wickens* in K.B. 29: 151, fig. 6 (1974). Type: Kenya, Kitui District, Mutha Hill, *Bally* 1656 (K, holo.!, EA, iso.!)

Shrub (?). Leaf-lamina elliptic, 1·3–2·7 cm. long, 0·5–1·6 cm. wide, very shortly and obtusely acuminate, base cuneate; petiole ± 0·5 mm. long. Inflorescence a 2–5-flowered solitary axillary pseudo-umbellate cyme; peduncle 2–4 mm. long; rhachis 2–5 mm. long; pedicels 5–7 mm. long. Calyx-tube obconical, 1·5 mm. long, 2 mm. in diameter; lobes shallowly deltoid. Petals ovate, 2 mm. long, 2 mm. wide, white. Anthers 1 mm. long; filaments 3 mm. long. Fruit not known.

KENYA. Kitui District: Mutha Hill, 25 Jan. 1942 & 25 Feb. 1942, *Bally* 1634! & 1656!
DISTR. **K4**; not known elsewhere
HAB. Evergreen mist forest; 1200 m.

SYN. *M. sp. nov.* sensu K.T.S.: 264 (1961)

NOTE. Related to *M. teitense* from **K7**.

22. **M. deminutum** *Brenan* in K.B. 1: 91, fig. 1/2 (1947); T.T.C.L.: 311 (1949); A. & R. Fernandes in Mem. Soc. Brot. 11: 60 (1956) & in Bol. Soc. Brot., sér. 2, 34: 198 (1960). Type: Tanzania, Pare District, Mtonto, *Greenway* 6530 (K, holo.!, EA, FHO, iso.)

Evergreen shrub up to 4·5 m. high. Leaf-lamina broadly rhombic-ovate to broadly elliptic, 0·5–1 cm. long, 0·4–0·9(–1·1) cm. wide, apex obtuse, base cuneate or sometimes rounded; petiole 0·5–1·5 mm. long. Inflorescence a 1–7- or more-flowered pedunculate axillary cyme; rhachis 1–4 mm. long, peduncle absent or 3·5–5 mm. long; pedicels ± 0·5 mm. long; bracts 1–1·5 mm. long. Calyx obconical, 1·3–1·5 mm. long, 1·8 mm. in diameter, shallowly sinuate with lobes 0·2 mm. high. Petals ovate-suborbicular, 1·4 mm. long, 1–1·3 mm. wide, white. Anthers 1·3–1·5 mm. long; filaments 2 mm. long. Fruit globose, 4–5 mm. in diameter.

TANZANIA. Lushoto District: Magamba–Gologolo, 6 Nov. 1947, *Brenan & Greenway* 8294! & Shagayu [Shagai] Forest Reserve, June 1951, *Eggeling* 6159! & Mkussu [Mkuzu] Forest Reserve, 22 Apr. 1969, *Yusufu* 5!
DISTR. **T3**; not known elsewhere
HAB. Upland rain-forest; 1800–2200 m.

23. **M. flavovirens** *Bak.* in K.B. 1897: 268 (1897); Gilg in E.M. 2: 44 (1898); V.E. 3 (2): 769 (1921); T.T.C.L.: 311 (1949); A. & R. Fernandes in Bol. Soc. Brot., sér. 2, 34: 201 (1960); F.F.N.R.: 310 (1962); A. & R. Fernandes in C.F.A. 4: 119 (1970). Type: Malawi, *Whyte* (K, holo.!)

Shrub or small tree up to 6 m. high. Leaf-lamina narrowly elliptic to obovate, 2·5–8 cm. long, 1·5–5·5 cm. wide, apex rounded or emarginate, base cuneate, yellowish green; petiole 0·5–2 mm. long. Inflorescences axillary few-flowered subsessile or shortly pedunculate racemose cymes; peduncles mostly 0–3(–6) mm. long, some sometimes longer; rhachis 1–9 mm. long; pedicels ± 2 mm. long. Calyx-tube obconical-campanulate, 2 mm. long, 2 mm. in diameter; lobes shallowly deltoid. Petals broadly ovate, 2 mm. long, 2 mm. wide, white. Anthers ± 1 mm. long; filaments 2·5 mm. long. Fruit globose, ± 15 mm. in diameter, black.

TANZANIA. Iringa District: Malangali, 18 Jan. 1941, *Lindeman* 1001!; Songea, 8 Nov. 1956, *Semsei* 2600!; Tunduru, 8 Nov. 1950, *Tanner* 220!
DISTR. **T7**, 8; also in Malawi, Zambia and Angola
HAB. *Brachystegia* woodland; 990–1500 m.

SYN. *M. cyanocarpum* Gilg in E.J. 30: 366 (1901); V.E. 3 (2): 769 (1921); T.T.C.L.: 311 (1949). Type: Tanzania, Iringa District, Mgololo, *Goetze* 763 (B, holo.†)

IMPERFECTLY KNOWN SPECIES

24. M. sp. A.

Tree 15 m. high. Leaf-lamina ovate to ovate-oblong, 6–11 cm. long, 3–6 cm. wide, abruptly acuminate, base cuneate, papillose on both surfaces; midrib impressed above, prominent beneath; petiole 5–8 mm. long. Inflorescence not known; pedicel (of fruit) 5 mm. long. Flowers not seen. Fruit globose, 16 mm. in diameter, black.

TANZANIA. Iringa District: Nyumbanitu, 6 Oct. 1958, *Ede* 67!
DISTR. **T7**; not known elsewhere
HAB. Upland rain-forest; 1500 m.

NOTE. Further collections, especially of flowering material, required. Near *M. viridifolium* Exell from which it is clearly distinguished by its habit, a 15 m. high tree as against a 3 m. high shrub, and in the larger fruits, 16 mm. in diameter, compared with 6 mm.

25. M. sp. B.

Shrub 3 m. high. Leaf-lamina ovate, 3–5 cm. long, 1·2–2 cm. wide, long-acuminate, base shortly cuneate to rounded, papillose on both surfaces; midrib impressed above, subprominent beneath; petiole 4–5 mm. long. Inflorescence 1–2-flowered; peduncle 4 mm. long; pedicel (of fruit) 2 mm. long. Flowers not seen. Fruit globose, ± 7 mm. in diameter, blackish red.

TANZANIA. Mpanda District: Kungwe Mt., 11 Sept. 1959, *Harley* 9597!
DISTR. **T4**; not known elsewhere
HAB. Upland rain-forest; 2220 m.

NOTE. Further collections, especially of flowering material, required. Near *M. fragrans* A. & R. Fernandes from which it may be distinguished by the longer petiole, 4–5 mm. as compared with 1–2 mm., and the longer pedicel of the fruit, 2 mm. instead of ± 0·5 mm.

26. M. sp. C.

Shrub 2–3 m. high. Leaf-lamina ovate-elliptic to broadly elliptic, 2·5–5 cm. long, 1·2–2·5 cm. wide, acuminate, cuneate to slightly rounded at the base, chartaceous, finely papillose; petiole up to 1 mm. long. Inflorescences 1–2 in the axils, 1(–2)-flowered; peduncle 1–3 mm. long; rhachis 1–3 mm. long; pedicels ± 0·1 mm. long; bracts triangular-acuminate, 1–1·5 mm. long. Calyx obconical, 1·8–2 mm. long, 2 mm. in diameter; lobes scarcely formed. Petals subtriangular, 1·5–2 mm. long, white. Fruit globose, 6–7 mm. in diameter.

TANZANIA. Morogoro District: Turiani Falls, 8 Dec. 1935, *B. D. Burtt* 5418! & Turiani, Nov. 1953, *Semsei* 1477! & Mtibwa Forest Reserve, Aug. 1952, *Semsei* 901!
DISTR. T6; known only from the Nguru Mts.
HAB. Rain-forest and riverine forest; ± 750 m.

SYN. [*M. verruculosum* sensu A. & R. Fernandes in Bol. Soc. Brot., sér. 2, 34: 87, 199 (1960), quoad specim. *Semsei* 901, *non* Brenan]

NOTE. The status and closest relationship of these plants is uncertain, but in the form of its leaves and inflorescences it is intermediate between 16, *M. verruculosum* and 17, *M. myrtilloides*.

INDEX TO MELASTOMATACEAE

GEOGRAPHICAL DIVISIONS OF THE FLORA

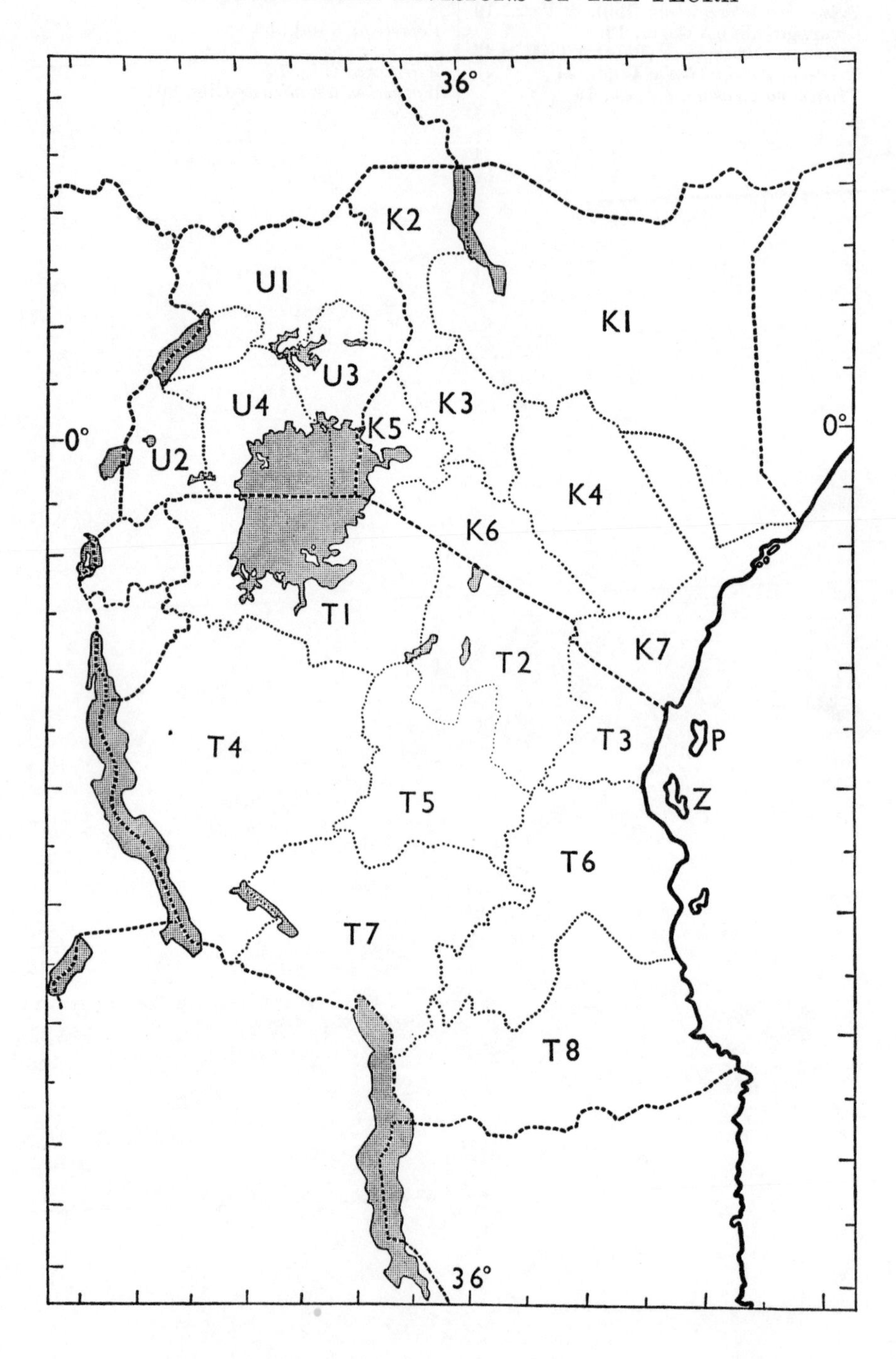